HOW's IT HANGING?

D0089770

HOW's IT HANGING?

EXPERT ANSWERS TO THE QUESTIONS MEN DON'T ALWAYS ASK

DR. NEIL BAUM AND DR. SCOTT MILLER

Skyhorse Publishing

Copyright © 2018 by Dr. Neil Baum and Dr. Scott Miller

All rights reserved. No part of this book may be reproduced in any manner without the express written consent of the publisher, except in the case of brief excerpts in critical reviews or articles. All inquiries should be addressed to Skyhorse Publishing, 307 West 36th Street, 11th Floor, New York, NY 10018.

Skyhorse Publishing books may be purchased in bulk at special discounts for sales promotion, corporate gifts, fund-raising, or educational purposes. Special editions can also be created to specifications. For details, contact the Special Sales Department, Skyhorse Publishing, 307 West 36th Street, 11th Floor, New York, NY 10018 or info@skyhorsepublishing.com.

Skyhorse® and Skyhorse Publishing® are registered trademarks of Skyhorse Publishing, Inc.®, a Delaware corporation.

Visit our website at www.skyhorsepublishing.com.

10 9 8 7 6 5 4 3 2 1

Library of Congress Cataloging-in-Publication Data is available on file.

Cover design by Rain Saukas

ISBN: 978-1-5107-2827-1
Ebook ISBN: 978-1-5107-2828-8

Printed in the United States of America

Contents

Introduction

For the past two decades, women have outlived their male counterparts by nearly five years. This has been an enigma to both of us, who have nearly sixty years of combined experience taking care of men. We have used this statistic as our motivation for writing this book: to help men live as long as women.

There have been many explanations for why women live longer than men. First of all, women continue to have medical care after they graduate from the pediatrician to the obstetrician/gynecologist during their reproductive years. They are advised to get regular Pap smears for cervical cancer and regular mammograms to check for breast cancer. Men, on the other hand, will seldom see a physician between their teenage years until they reach middle age, around age fifty. Historically, men have an attitude of "if it ain't broke, don't fix it!" There is often a reluctance among men to go to their doctors for routine medical examinations, wellness checks, or for emotional challenges. Men often talk about spending more time caring for their cars or planning for a vacation than looking after their health. Men often place work before health care needs as well. We want this book to serve as a guideline for men to give them information and knowledge about their health and how to address and overcome their unique health care problems. Making men more aware of their health is one of our primary motivations for writing this book.

Unfortunately, men lack awareness of and understanding about their health care needs. Men are also historically reluctant to seek health care in a timely fashion and may have a decreased ability to communicate their feelings and emotions. Men are also more likely to make risky lifestyle choices. For example, 40 percent of men smoke compared to 25 percent of women. Men also consume more alcohol than women, are twice as like to be overweight as women, and represent the majority of drug addicts. Society appears to take a passive attitude when our celebrity icons succumb to a tragic death from lifestyles associated with sex, drugs, and rock and roll. Recent examples include Jimi Hendrix, James Dean, Robin Williams, and Otis Redding, just to name a few.

Men also suffer from unique cancers, including prostate and testicular cancers. Prostate cancer is the second most common cause of cancer death in men, resulting in thirty thousand deaths in the U.S. each year, and rising steadily. It is the hope that this book will motivate men to have annual testing for prostate cancer.

Testicular tumors have a peak incidence around age thirty and are the most common tumors in men between ages twenty-four to thirty-four years. As with prostate cancer, early diagnosis is essential, and with modern treatments, the mortality from testicular cancer continues to fall.

With the aging of the baby boomers, there will be millions of middle age and older men who will be impacted by male hormone testosterone deficiency, a counterpart but very rarely discussed and even more rarely understood life phase to menopause among women. This book will discuss the symptoms of andropause and what treatments are available to help nearly every man with testosterone deficiency.

Our goal is to offer men all the information that they need to make good decisions regarding their health. Perhaps after reading this book and taking action on our recommendations, we can narrow the longevity gap between men and women. We believe all men *and* women are created equal and men deserve to have the same life expectancy as women.

Dr. Scott Miller
Dr. Neil Baum

CHAPTER 1

Anatomy and Physiology Down There

Most men know very little about what is going on down there, how to keep things working, and when things are not quite right. Most men will direct their focus on the penis, with little attention paid to the sack hanging in the background. Let's begin our journey by exploring these treasured jewels and all of the related parts of the total package making up the male machine.[1]

Penis—the Helmeted Soldier

The penis has several specialized parts, each with its own function. However, all of these parts work in concert to provide erections, to facilitate release of semen during ejaculation, and to allow passage of urine. The amazing male anatomy allows

---

1. Campbell, Meredith Fairfax, Patrick C. Walsh, Alan J. Wein, and Louis R. Kavoussi. *Campbell-Walsh Urology*. Philadelphia, PA: Elsevier, 2016.

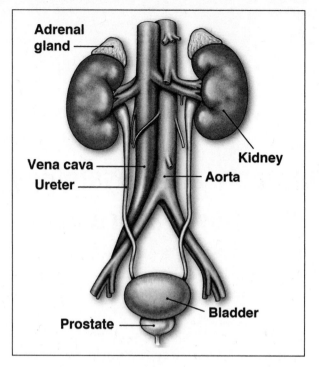

Figure 1: Male Parts *(Boyter)*

ejaculation and urination to use the same pipes and plumbing, but it is so beautifully organized that there is no mixing of urine and semen.

Urethra

The urethra is a tube or a pipe that provides a passageway for urine to exit the bladder. The male urethra has three sections. The first section, closest to the bladder, is the prostatic urethra. In this portion, the urine passes from the bladder through the middle of the prostate gland (discussed later in this chapter). The second section is the membranous urethra. This portion

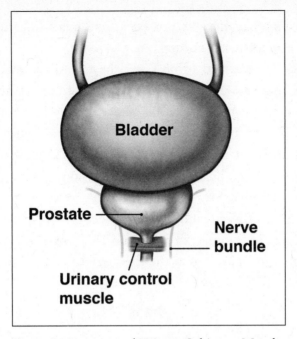

Figure 2: Prostate and Urinary Sphincter Muscle
(C. Boyter)

passes through the muscular pelvic floor and urinary sphincter muscle. The membranous urethra provides most of the urinary control in order to prevent leakage (incontinence). The final and longest section is the pendulous, or penile, urethra.

The urethra contains small glands that keep the lining moist. During arousal, these glands produce additional fluid (often called "pre-cum") in order to pave the way for the ejaculation to follow. These glands can also produce additional fluid when irritated, such as with an infection. This overproduction results in a discharge from the urethra.

The final opening of the urethra is called the meatus. It is usually located at the very tip of the penis, but on occasion can be located on the undersurface. This condition is called hypospadias and can cause a misdirected urinary stream or difficulty achieving a pregnancy. On rare occasions, a false meatus can be located just above the normal one.

When it comes to urine flow, the urethra is much more complex than a simple hose. The portion of the urethra in the middle of the head of the penis has a slightly larger diameter than the immediately adjacent portions. This gentle disturbance in flow causes the exiting urine to spiral into a focused stream.

Narrowing of the urethra at any location can cause serious urinary difficulty. Scar tissue formation, referred to as stricture, can occur as a result of prior infection, surgery (urethral or prostate), or trauma. Meatal stenosis is a condition in which the narrowing is located at the final opening of the urethra.

Glans

The glans of the penis is most commonly referred to as the head of the penis. It is shaped like a soldier's helmet, thereby allowing smooth penetration during sexual intercourse. Its structure is very different than the remainder of the penis.

The skin of the glans is very unique. Although the glans expands during an erection, the skin does not stretch nearly as

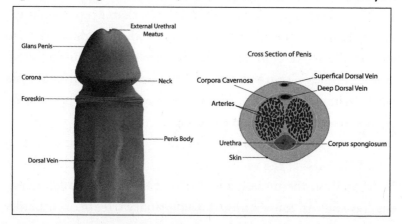

Figure 3: The Erection Mechanism *(Shutterstock)*

well as skin on other parts of the body. The glans is richly supplied with blood vessels and all types of sensory nerves (light touch, temperature, pressure, pain).

Below the skin of the glans, a specialized structure—the corpus spongiosum—fills with blood during an erection. The corpus spongiosum originates in the pelvis and travels as a narrow tube surrounding the urethra until it reaches the glans.

Shaft

Unlike the glans, the shaft of the penis is surrounded by very elastic skin. It is generally hairless, except for the portion closest to the body. The shaft's skin is richly supplied with nerves, but most of these are sensitive to light touch.

The penile shaft contains two large cylinders, one on each side. These cylinders—each referred to as a corpus cavernosum—fill with blood during an erection. The outer lining of these cylinders is a tough and non-expansible layer called the tunica albuginea. Since this outer layer does not expand, once the corpus cavernosum reaches capacity, the penis will become hard and rigid in order to allow penetration at the time of sexual intimacy.

The corpus cavernosum is controlled by a very special nerve, aptly named the cavernosal nerve. This nerve travels along both sides of the prostate prior to exiting the body and reaching the penis. The cavernosal nerve is solely responsible for initiating an erection; it has no direct role in penile sensation or sexual climax.

Foreskin

In an uncircumcised male, the foreskin is a hood of redundant skin that covers the head or glans of the penis. During an erection, the glans usually protrudes from the foreskin. When circumcised, the redundant skin is surgically removed, and leaves the glans exposed.

Scrotum—the Sack of Jewels

How is the scrotum like a woman's purse? It contains some valuables and many things that are hard to find.

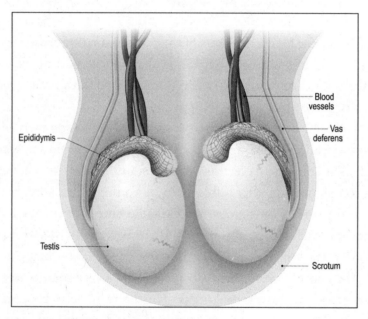

Figure 4: The Scrotum *(iStock)*

Testicles

Although men do not pay much attention to their testicles, even a minor trauma will bring them to the forefront. Behind the scenes, a variety of specialized cells provide some essential functions. Germ cells produce sperm. Leydig cells produce testosterone. Other cells, such as Sertoli cells, serve more of a supporting role. A complex network of small tubes—the seminiferous tubules—travel throughout the testicles, allowing the sperm to reach their next destination: the epididymis.

Epididymis

Sperm travel through the epididymis during their final maturing process. The epididymis is located behind each testicle and can feel lumpy and irregular during monthly self-examination. The epididymis should also be easily distinguishable from the testicle itself. A lump in the epididymis is seldom concerning, whereas a lump on the testicle should raise the concern of testicular cancer.

Vas deferens

The vas deferens is a very long tube that carries sperm from the epididymis to the urethra. It takes a circuitous route from the scrotum, deep into the groin, through the pelvis, behind the bladder, and into the urethra near the prostate. It is this tube that is divided through a small opening in the scrotum during a vasectomy, thereby providing an effective means of permanent birth control. Since less than 5 percent of the total volume of semen comes from the testicles, no change is noticed during ejaculation following a vasectomy (see Chapter 6).[2]

Spermatic cord

The rope-like structure that connects the testicle to the body is called the spermatic cord. It contains the blood vessels and

---

2. "What Is in Semen?" New Health Guide. November 10, 2013. Accessed August 01, 2017. http://www.newhealthguide.org/What-Is-In-Semen.html.

nerves that supply the testicle, along with the vas deferens. The spermatic cord also contains a muscle, the cremasteric muscle. This controls the elevation of the testicles in response to decrease in external temperature and other types of stimulation.

Scrotal wall

The scrotal wall also has a muscle, the dartos muscle. It contracts in response to cold and contributes to the wrinkling effect on the scrotal skin.

Prostate

The prostate, along with two little adjacent structures called the seminal vesicles, provide only one function: to manufacture

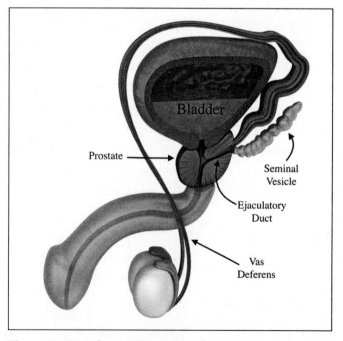

Figure 5: Ejaculatory Duct *(iStock)*

the majority of the fluid that makes up the semen. The prostate and seminal vesicles converge with the vas deferens at the urethra, thereby forming the ejaculatory duct. A mixture of all three fluids, to include the sperm from the testicles, enters the urethra at this location in order to deliver the semen in its final form to the outside world.

Although the prostate does not provide any other role, it is intimately surrounded by delicate structures that supply some very important functions. These include the urinary sphincter muscle that supplies urinary control (continence) and the cavernosal nerves that initiate an erection. As a result, treatment of prostate conditions can cause damage to these adjacent structures, along with a corresponding decrease in erectile function or development of incontinence (urinary leakage).

In addition, the urethra travels through the prostate as it exits from the bladder. As a result, enlargement, infection, and surgery of the prostate can restrict the flow of urine. These conditions will be discussed in detail.

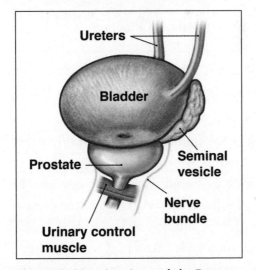

Figure 6: Hanging Around the Prostate
(*C. Boyter*)

Bladder

The bladder serves two roles—to store and to empty the urine produced by the kidneys. We tend to take these very complex functions for granted when all is working well. However, nerve damage or obstruction downstream can wreak havoc on bladder function, resulting in some serious quality-of-life issues.

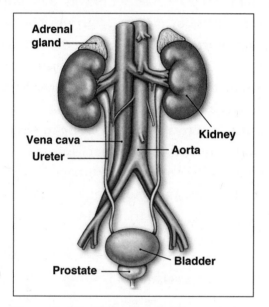

Figure 7: The Male Urinary Tract *(C. Boyter)*

Kidneys and ureters

The kidneys are amazing organs that affect every system of the body. Their primary functions are to filter our blood and to produce urine or liquid waste. Not only do they eliminate toxins and excess fluid, they also finely tune our body's chemistries, such as sodium and potassium levels.

The ureter—not to be confused with urethra—is the delicate tube that carries the urine from the kidney to the bladder.

It is also the tube that, when blocked by the passage of a small kidney stone, can cause severe enough back pain to bring a grown man to tears.

Adrenal Gland

The adrenal gland is a small gland located on the top of each kidney. Other than being its neighbor, this gland has no direct relationship to the kidney. In addition to producing the well-known adrenaline, it releases hormones that control steroid levels, salt levels, and other bodily functions in conjunction with other hormone-producing organs. It also produces a small amount of sexual hormones. Interestingly, cholesterol is the major component used to manufacture adrenal hormones.

Nerves

The structures down there have a variety of nerves to control their functions. These nerves separately control erections, sexual climax, skin sensation, urinary control (continence), and bladder emptying. Damage to these nerves can cause loss of any of the above functions.

Pituitary gland—the Other Control Center

The brain is often referred to as the largest sex organ. Although the brain deserves most of the credit for controlling our sexual response, a small gland located just below the brain—the pituitary gland—releases hormones under the direction of the brain (via the hypothalamus). One of these hormones—luteinizing hormone (LH)—controls the production of testosterone in the testicle.

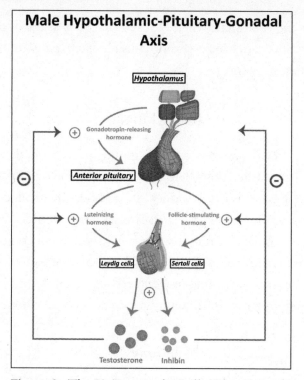

Figure 8: The Pit Bosses, the Balls *(Shutterstock)*

The pituitary gland produces a vast array of hormones that control a variety of bodily functions, both down there and elsewhere. Follicle stimulating hormone (FSH) stimulates the production of sperm in the testicles. Hormones from the pituitary gland also help control the function of the kidneys, adrenal glands, thyroid, and breasts.

Conclusion

Now that we know the basic makeup of everything hanging down there, let's delve into how to keep things running smoothly and fine-tuned, just like your automobile.

CHAPTER 2

Benign Prostate Conditions— When That Walnut is Acting a Little Nutty

Peter, age fifty-six, was feeling a bit feverish over the last few hours and knew something was not right down there. He had noticed some vague urinary and groin symptoms over the previous weeks that would "come and go" (not in the desirable sense of these words). He was referred to a urologist who diagnosed him with a prostate infection.

What Is the Prostate?

Most men do not know what a prostate is—that is, not until their prostate starts causing problems. The prostate is a small gland with the sole purpose of producing a portion of the fluid expelled during ejaculation or at the time of orgasm. The remainder of this fluid—or ejaculate—is produced by the seminal vesicles (two adjacent glands) and the testicles. The prostate does not provide any other sexual function. However, this small gland is surrounded by some very important and somewhat delicate structures that are responsible for normal urinary and sexual function.

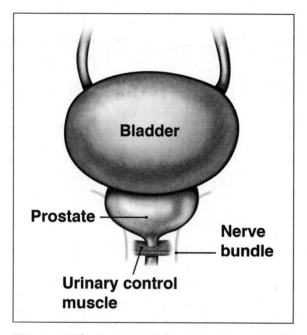

Figure 1: The Prostate and its Neighbors *(C. Boyter)*

The prostate resides below the bladder and just in front of the rectum. The bladder empties its contents of urine through a tube (the urethra) which passes through the center of the prostate. As a result, any problem with the prostate can cause a problem with urinary flow. Although the prostate can block urinary flow, it is the urinary sphincter muscle just below the prostate that prevents urinary leakage. In addition, the nerves responsible for erectile dysfunction travel along the undersurface of the prostate, adjacent to the rectum. Interestingly, the nerves responsible for orgasm and ejaculation are located deeper in the pelvis and are usually out of harm's way from prostate problems (and the treatments for these problems).

Prostate Enlargement—Does Size Really Matter?

The prostate increases in size with time, peaking in the late fifties. Benign prostatic hyperplasia (BPH) and bladder outlet obstruction (BOO) are other common terms for this common condition that occurs in middle-age and older men. As the prostate grows, it compresses or squeezes the urethra and impedes urinary flow. So size does matter? Yes and no. The consistency or firmness of the prostate can also affect how well the prostate "opens up" to accommodate the flow. In Peter's case, he probably had some degree of prostate enlargement that was exacerbated by "swelling" from the prostate infection.

When the prostate first begins to enlarge, it usually causes no urinary symptoms. At first, this enlargement does not cause any significant obstruction or blockage of the urine flow. The bladder is strong enough to overcome mild obstruction without causing noticeable symptoms. Once the obstruction becomes more significant, the first noticeable symptom is usually frequent urination rather than a reduced flow. The explanation for the urinary frequency is that the bladder becomes less efficient at emptying and it contracts more frequently. Consider this analogy. If you had a pile of bricks on one side of your house and wanted to move them to the other side, it would take quite a few trips if you were to carry them by hand one at a time. However, if you had a wheelbarrow, it would be much more efficient to place a number of bricks in the wheelbarrow, and you would make fewer trips moving the bricks. The same applies to your bladder that is trying to move a large quantity of urine; instead of a single contraction to empty the bladder, the bladder makes multiple contractions and empties only small quantities of urine with each contraction.

Treatment Options—Cracking the Walnut

The treatment approach to prostate enlargement is initially based on how bothersome these symptoms become. In addition to increasing urinary frequency both during the day and at night, prostate enlargement can cause restricted flow, straining to urinate, difficulty starting a stream, an increased urge to urinate, and dribbling of urine after urination. These symptoms

(Circle One Number on Each Line)	Not at All	Less Than 1 Time in 5	LessThan Half the Time	About Half the Time	More Than Half the Time	Almost Always
Over the past month or so, how often have you had a sensation of not emptying your bladder completely after you finished urinating?	0	1	2	3	4	5
During the past month or so, how often have you had to urinate again less than two hours after you finished urinating?	0	1	2	3	4	5
During the past month or so, how often have you found you stopped and started again several times when you urinated?	0	1	2	3	4	5
During the past month or so, how often have you found it difficult to postpone urination?	0	1	2	3	4	5
During the past month or so, how often have you had a weak urinary stream?	0	1	2	3	4	5
During the past month or so, how often have you had to push or strain to begin urination?	0	1	2	3	4	5
	None	1 Time	2 Times	3 Times	4 Times	5 or More Times
Over the past month, how many times per night did you most typically get up to urinate from the time you went to bed at night until the time you got up in the morning?	0	1	2	3	4	5

Add the score for each number above and write the total in the space to the right.   TOTAL: _____

SYMPTOM SCORE: 1-7 (Mild)   8-19 (Moderate)   20-35 (Severe)

Figure 2: AUA Prostate Symptom Score.

can be quantitated with the American Urological Association Symptom Score questionnaire (AUASS, see Figure 2).[1]

Note that an additional question assesses level of patient "bother" (Figure 3). This becomes important when speaking to your doctor as he\she will want to know how much discomfort or bother the symptoms are and how the symptoms are impacting your quality of life.

QUALITY OF LIFE (QOL)							
	Delighted	Pleased	Mostly Satisfied	Mixed	Mostly Dissatisfied	Unhappy	Terrible
How would you feel if you had to live with your urinary condition the way it is now, no better, no worse, for the rest of your life?	0	1	2	3	4	5	6

Figure 3: Urinary Bother Score

Lifestyle Changes

At low levels of symptoms, observation and lifestyle changes are most appropriate. For instance, reducing the amount of fluids consumed in the evening may prevent awakening so often at night to urinate. Diuretics and other medications that increase the volume of urine excreted by the kidneys, and thus delivered to the bladder, can be taken earlier in the day to reduce nighttime urination. In other cases, dietary modification such as decreasing caffeinated and alcoholic beverages—both of which can act like diuretic medications—will reduce the symptoms of prostate enlargement.

---

1. Barry, Michael J., William O. Williford, Yuchiao Chang, Madeline Machi, Karen M. Jones, Elizabeth Walker-Corkery, and Herbert Lepor. "Benign Prostatic Hyperplasia Specific Health Status Measures in Clinical Research." *The Journal of Urology*, 1995, 1770–774. doi:10.1097/00005392-199511000-00051.

A Simple Pill May Do

When symptoms become more bothersome, a variety of medications are available to relax or shrink the prostate. Alpha-blockers are the most common type of medication to treat prostate enlargement. Examples include Flomax® (tamsulosin), Uroxatral® (alfuzosin), and Rapaflo® (silodosin). These alpha-blocker medications work by relaxing the muscles in the prostate gland and thus increase the flow of urine from the bladder through the urethra to the outside of the body. Both the prostate and the bladder that connects to the urethra contain an abundance of alpha nerve receptors that, when blocked by these medications, cause the bladder opening to relax and open.

Another class of medications, five-alpha reductase inhibitors, actually cause the prostate to shrink in size. Examples include Proscar®(finasteride) and Avodart®(dutasteride). These inhibitors tend to be more effective when the prostate is significantly enlarged, perhaps as large as a small peach or plum. Both Proscar® and Avodart® may require four to six months before prostate gland decreases in size. It is not uncommon for your doctor to prescribe both alpha-blockers combined with either Proscar® or Avodart®. In fact, one drug, Jalyn®, combines both of these medications into one pill.

An abundance of prostate supplements exists. The most common supplement is saw palmetto. However, none of these supplements have been scientifically shown to be effective in comparison to placebo (see Chapter 13).

When Pills Just Won't Do the Job

If the medications become ineffective or poorly tolerated, a surgical option to treat the prostate enlargement can be the best course of action. A surgical approach is also the best option when the prostate enlargement causes other problems such as recurrent infections, blood in the urine, bladder stones, bladder damage, kidney damage, or significant urinary retention. The traditional surgery for relieving prostate obstruction is a trans-urethral resection of the prostate, also referred to as a TURP. During this procedure under anesthesia, the urologist places a telescope through the urethra or the tube in the penis and into the opening of the prostate. Using a specialized electric knife, the urologist removes the inner portion of the prostate in small pieces leaving the outer portion of the prostate intact.

This would be similar to remove the fruit in the orange and leave the peel intact. This creates a large channel through which the urine can easily pass, thus decreasing the blockage of urine. Although this is a well-tolerated procedure, it is occasionally associated with side effects. A large percentage of patients experience retrograde ejaculation in which the ejaculate goes into the bladder rather

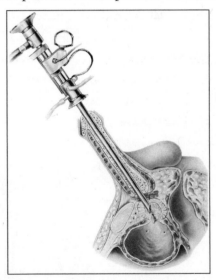

Figure 4: Transurethral Resection of the Prostate (TURP) *(Shutterstock)*

than out of the penis during orgasm (see Chapter 5).[2] Also, a

2. Campbell, Meredith Fairfax, Patrick C. Walsh, Alan J. Wein, and Louis R. Kavoussi. *Campbell-Walsh Urology.* Philadelphia, PA: Elsevier, 2016.

few patients can experience bleeding, urinary incontinence, or erectile dysfunction.

There are several minimally-invasive procedures that have been developed as an alternative to a TURP in order to reduce recovery and to decrease the possible side effects. For instance, several iterations of lasers have been used to replace the electric knife used during a TURP. The most common "laser-TURP" used today is the GreenLight™ procedure. Other types of heat can be used to destroy, shrink, or soften portions of the prostate—often with just local anesthesia and in some instances can be accomplished in the doctor's office. Examples include transurethral radiofrequency ablation (TUNA®), transurethral microwave therapy (TUMT, Prolieve®), and water vapor (steam) therapy (Rezūm®). A more recent novel technique—the UroLift® procedure—uses small "drawstring-like" implants that your urologist places telescopically inside the prostate to pull open the tissue.

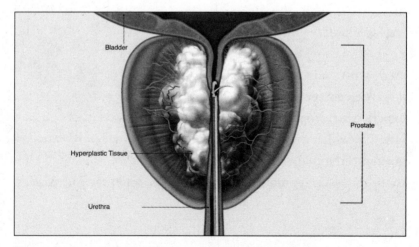

Figure 5: Rezūm® Treatment *(courtesy of NxThera)*

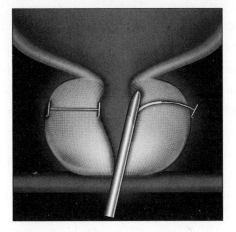

Figure 6: UroLift® *(Neotract)*

On rare occasions, the prostate is too large to treat through the urethra. In these cases, the inner portion of the prostate or the part of the prostate that is obstructing the flow of urine is removed through an abdominal incision. This procedure is similar to prostate removal for cancer; however, for benign conditions only the inner portion is removed, thereby lessening the risk for urinary or sexual side effects. Laparoscopic or robotic techniques can also be used.

How to Determine Which Treatment Is Right for You

The need for and type of treatment recommended by the urologist is usually based on symptoms. However, your urologist may want to perform some other tests. A urinalysis will be performed as a screening test for other potential causes of your symptoms such as a urinary tract infection, kidney or bladder stones, blood in the urine, undiagnosed diabetes, or (in rare cases) bladder cancer. Although prostate cancer can mimic the same conditions mentioned in this chapter, that is rarely the case. In fact, even for someone who has been diagnosed with prostate cancer, any urinary symptoms are more likely to be related to coexisting benign prostate enlargement. Nevertheless, your urologist may obtain a PSA (prostate-specific antigen) blood test to screen for prostate cancer. Also, a baseline PSA blood test is helpful since levels could be falsely elevated by some of the treatments for prostate enlargement (see Chapter 3).

Although not usually necessary with mild symptoms, your urologist may also order some diagnostic tests to assess the severity of your condition. As a general rule, patients with other conditions such as diabetes, kidney disease, or neurologic disease will require more in-depth evaluation and earlier active treatment. Among the simpler tests is a bladder ultrasound performed shortly after you empty your bladder (post-void residual, PVR). This easy and noninvasive test will determine how completely your bladder empties—and how much it retains. Patients are often surprised by how much urine their bladder retains in the absence of severe symptoms. Unless other conditions are suspected, CT scan, MRI, and kidney ultrasound are seldom necessary. If the urinary frequency or urgency is thought to be a result of a bladder condition in addition to or instead of a prostate condition, urodynamic testing may be needed (see "Overactive Bladder" later in this chapter). Urodynamic testing involves placing a very slender tube through the urethra into the bladder. The tube has a small pressure sensor at the tip. A computer then slowly fills the bladder and plots the internal bladder pressure and bladder volumes throughout filling and emptying. The computer also plots the flow rate and measures the retained volume.

Prostate Infection—Knocking On the Back Door

Prostate infections come in many forms. Since the infection usually causes the prostate to "swell," symptoms can include all of those mentioned for prostate enlargement (urinary frequency both during the day and at night, restricted flow, urinary straining, difficulty starting a stream, an increased urge to urinate, and urinary dribbling). However, with infection, the symptoms usually develop over days to weeks—rather than months to years as with benign enlargement. A prostate infection can also cause pain with urination (dysuria) or generalized pain in the groin, lower abdomen, testicles, lower back, and even the legs. On occasion, as in Peter's case, an infection can cause fever and chills.

Unlike benign prostate enlargement, a prostate infection always requires treatment. Evaluation is usually limited to physical examination, urinalysis, and urine culture. More involved testing such as that mentioned with prostate enlargement is usually reserved for cases in which other conditions are suspected or when a patient does not respond to treatment. Finding the offending organism can often be a challenge since the prostate can harbor bacteria without allowing them to escape into the urine. Antibiotics are usually targeted towards the most likely organism to cause urinary tract infections (or in some cases, sexually transmitted infections—see Chapter 12). Since it is difficult for antibiotics to penetrate the prostate due to its poor blood supply, it can take four or more weeks of antibiotics to eradicate the culprit. As always, finish your antibiotic as prescribed by your physician. Our take-home message is that it is imperative that you follow the doctor's instructions and continue with the medication even after your symptoms have subsided.

In order to facilitate the antibiotic therapy, your urologist may also prescribe alpha-blockers (as used with benign prostate enlargement) to relieve symptoms and reduce the retention of any infected urine. Increased fluid intake will also help "wash out" or dilute any bacteria in the urinary tract. Anti-inflammatory medication such as ibuprofen and urinary analgesics such as pyridium can help relieve symptoms of pain and burning associated with a prostate infection.

Once the symptoms have resolved, your urologist may evaluate you for any risk factors that would lead to a prostate infection. The most common risk factor is actually benign prostate enlargement, as is probably the case for our dear Peter. When some prostate enlargement symptoms remain after completely treating the infection, your urologist may recommend continuing alpha-blockers in order to prevent recurrence of the infection.

On occasion, the infection can be difficult to eliminate fully. Sometimes the solution can be as simple as an extended course of the same or different antibiotic. When standard cultures do not reveal the offending organism, prostate secretions can be obtained using a prostate massage in order to send for culture. In the case of sexually transmitted infections, the partner must also be treated—even if your partner has no symptoms. When these measures are unsuccessful, ultrasound, CT scan, urodynamic testing, and/or cystoscopy may be needed to look for hidden causes.

When patients experience repeated and/or prolonged prostate infections, the condition can become more chronic in nature. "Chronic prostatitis" is usually caused by inflammation rather than infection—although a bacterial infection was the inciting event. As a result, seldom does chronic prostatitis respond to antibiotics. Anti-inflammatory medication (ibuprofen) and alpha-blockers are the mainstay therapy. In refractory cases, minimally invasive therapies for prostate enlargement such as laser vaporization, microwave, or UroLift have been tried. At the present, there is no scientific evidence that minimally-invasive treatments are effective in treating chronic prostatitis.

Pain in the . . . Prostate

Prostate pain syndromes can be the biggest challenge for a urologist and the most aggravating for the man suffering from this malady to evaluate and treat. Many varieties exist, including "prostadynia" (isolated prostate pain), pelvic pain syndrome, and interstitial cystitis. Evaluation begins with eliminating the conditions already discussed in this chapter. An accurate diagnosis is important since this condition seldom responds to surgical management.

Treatments for prostate pain syndromes are directed by the symptoms themselves. As mentioned earlier in this chapter, alpha-blockers relax the prostate. However, in some cases, it is the muscles around the prostate that experience "spasm." For these patients, anti-spasmodics such as valium may be helpful. In other cases, a nerve stimulator (Interstim®) can be implanted in the tailbone "to distract" the nerves leading to the bladder, prostate, and pelvis. These conditions and treatments are discussed in detail in Chapter 10.

Overactive Bladder— The Prostate's Menacing Neighbor

Although the most common cause of urinary frequency in a middle-aged male is prostate enlargement, the bladder can also be the culprit. A bladder is like a crying baby. Whether a baby has a dirty diaper, is hungry, or is just tired, the cry sounds the same. Similarly, an overactive bladder, an enlarged prostate, a bladder infection, or some combination of these conditions can all lead to frequent urination. As such, in a man with an enlarged prostate whose main symptoms are urinary frequency and urgency, but whose bladder does not retain significant urine, treatment might be better directed towards "calming" the bladder with anticholinergic medications (Ditropan®, Detrol®, Toviaz®, Vesicare®) or bladder-specific beta agonist medication (Merbetriq®). However, these "bladder-calming" medications can work too well and cause your bladder to retain urine (especially with coexisting prostate enlargement). Therefore, your urologist may repeat the ultrasound measurement of your bladder immediately following urination (post-void residual, PVR) once you have been on this therapy for several weeks. Other possible side effects of anticholinergic medications include dry mouth, constipation, blurred vision, and occasionally confusion (particularly in the elderly). Since Myrbetriq® works through a different mechanism, these side effects can be avoided. However, Myrbetriq® can cause an elevation of blood pressure in a minority of patients.

So What Happened to Peter?

Peter was placed on a common antibiotic (ciprofloxacin), Flomax, and ibuprofen. Within one week his symptoms resolved, but he completed the antibiotics as prescribed by his doctor. He no longer needed the ibuprofen after the first week, but to this day he continues using the Flomax. His flow has never been better!

Bottom Line

Prostate ailments can present in a variety of forms. As such, your doctor will evaluate your specific condition and will guide you through the best treatment options. Early diagnosis and treatment are the key to avoiding a chronic prostate inflammatory or pain syndrome. The prostate is man's best friend during his youth and fertility years. Upon reaching middle age, it is a gland of pain and discomfort. However, there are effective treatments so that men don't have to suffer from conditions "down there" related to his prostate gland.

CHAPTER 3

Prostate Cancer—Fingering the Little Culprit

Jerry just started his new job at a large investment banking firm. At age forty-nine he was the picture of health, but he still decided to take advantage of the executive physical offered as benefit by his company. He passed with flying colors except for one abnormal blood test—a slightly elevated PSA (prostate-specific antigen) blood test. Suddenly, the possibility of prostate cancer was looming in front of him.

The company physician referred him to a urologist, Dr. Samples. In the weeks to follow, the urologist repeated the PSA test and ordered additional blood tests. Eventually, Jerry underwent a prostate biopsy. Unfortunately, the biopsy revealed prostate cancer. Dr. Samples sat down with Jerry and his wife to formulate the best treatment for Jerry.

Introduction

The prostate gland is one of those organs that most men give very little consideration until something goes wrong. Prostate issues can manifest in an array of symptoms, but the prostate can stealthily reveal no symptoms—or even confusing symptoms—when it misbehaves. The most common prostate problems are enlargement, infection, and cancer. In this chapter, we will focus on everything you need to know about prostate cancer (see Chapter 2 for information about other prostate concerns).

What Is the Prostate Gland?

The prostate gland is a small, walnut sized, rubbery organ that is located just below the male bladder. It resides within an inch of the skin located between the scrotum and anus (the medical term for this external area is the "perineum"). Adding to its cryptic nature, it surrounds the urinary tube (urethra) as it exits the bladder. Since you urinate through a passageway in your prostate, any problems with this gland can significantly affect your urine flow and produce bothersome urinary symptoms.[1]

The primary function of the prostate gland is to produce some of the fluid for ejaculation that is mixed with fluid from the neighboring seminal vesicles and sperm produced in the testicles. All three of these structures join together as one common tube (the ejaculatory duct, see Figure 1) just prior to entering the urethra.

---

1   Campbell, Meredith Fairfax, Patrick C. Walsh, Alan J. Wein, and Louis R. Kavoussi. *Campbell-Walsh Urology*. Philadelphia, PA: Elsevier, 2016.

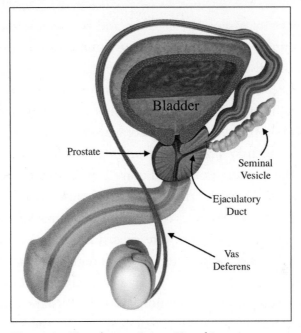

Figure 1: Ejaculatory Duct *(iStock)*

The prostate has no other urinary or sexual function. However, it is intimately surrounded by delicate nerves and organs involved in urinary, sexual, and bowel function. Also, the prostate gland does not produce any male hormones such as testosterone. These are primarily produced by the testicles.

So, What Is Prostate Cancer?

A cancer forms when the genetic material, or DNA, inside a cell is altered in such a way that the cell can no longer follow the organizational rules of its neighboring cells. This rogue cell therefore has the capacity to invade nearby tissues, as well as far away organs. However, this change in DNA—also referred to as a "mutation"—must be minimal enough so as to preserve its essential functions and to avoid detection by your immune system. In fact, most mutations result in the death of a cell.

Prostate cancer is the second most common cancer in men (behind skin cancer), perhaps as a result of the proportionately low blood supply in the prostate compared to other organs. This characteristic may protect it from exposure to your immune system (the "cancer police" who are responsible for removing any abnormal cancer cells before they cause trouble). Think of the prostate gland as a very crowded city packed with citizens—or in this case, prostate cells. Like any overpopulated area, the "cancer police" (or your immune system) struggles to enforce the rules. On the other hand, once a cell escapes, it must contend with your immune system. Perhaps this formidable immunologic ability helps explain why most men with prostate cancer survive the disease. Nevertheless, since prostate cancer is so common, it is the second leading cause of cancer death in men, with lung cancer being the most common cause of death in middle age and older men.

Uncovering the "Terror Cell"

Most patients with prostate cancer do not have any symptoms at the time of diagnosis. Early prostate cancer typically has no symptoms. Once the cancer starts causing symptoms, it has probably already spread beyond the prostate gland. Therefore, screening is the most common method of detecting prostate cancer. Two tests are commonly used for this detection. One is a blood test called prostate-specific antigen, or PSA. The other is a digital rectal exam (DRE).

To Screen or Not to Screen, That Is the Question

A lot of controversy surrounding prostate cancer screening exists. Recommendations from a variety of large organizations are all over the board, from no screening at all to screening all men yearly starting at age 40. Like most controversies, the answer probably resides somewhere in the middle.[2]

In a nutshell, concerns about aggressive screening include the risks of unnecessary additional testing, overtreatment of individuals who might do well with less or no treatment, and side effects from testing or treatment. On the other hand, in the absence of screening, most prostate cancers would be discovered when a man develops symptoms. This delay would most likely deny a man a reasonable chance for cure, since prostate cancer has usually escaped the prostate once symptoms develop.

So, what is the take-home message? Starting at age forty, a man should learn about their own particular risk of developing prostate cancer compared to the risks of screening.[3] Besides, this is a good time to start seeing a physician for routine health-maintenance exams. For men with a family history of prostate cancer, such as a father, brother, or uncle who have been diagnosed with prostate cancer, or African-American men, who have a slightly higher increase in prostate cancer than occurs in Caucasian males, early screening is probably beneficial. A single baseline PSA blood test at an early age can also help determine future risk by serving as a baseline. In the absence

---

2. "Draft Recommendation Statement." Prostate Cancer: Screening—US Preventive Services Task Force. Accessed August 01, 2017. https://www.uspreventiveservicestaskforce.org/Page/Document/draft-recommendation-statement/prostate-cancer-screening1.

3. "Educate." ProstAware. Accessed August 01, 2017. http://prostaware.org/educate/.

of an alternative method for detecting prostate cancer at a curable stage, every man between the ages of forty and seventy-five should have an individualized screening plan. This plan can be created with a careful discussion with your doctor.[4]

Diagnosing Prostate Cancer

An abnormal PSA blood test or prostate exam does not necessarily mean that you have prostate cancer. This blood test can be elevated for other reasons, such as a prostate infection. Recent ejaculation and long distance bike riding can also irritate the prostate enough to raise the PSA level. An individual man's level can vary based on age and prostate size. In order to make a diagnosis, a prostate biopsy is necessary. However, the following simple tests can help determine when PSA levels are elevated for reasons other than cancer, potentially allowing a man with an elevated total PSA to avoid a biopsy:

- Free PSA—Prostate-specific Antigen is a protein that travels in the bloodstream in different forms, often attached to other proteins. One of these forms— "free-PSA"—floats around unattached to other proteins. Higher free-PSA levels are correlated with the likelihood of non-cancerous causes for the rise in total PSA levels. Accordingly, free-PSA is sometimes referred to as "good PSA," similar to "good cholesterol" (HDL). The free PSA is often expressed

---

4. "American Cancer Society Recommendations for Prostate Cancer Early Detection." American Cancer Society. Accessed August 01, 2017. https://www.cancer.org/cancer/prostate-cancer/early-detection/acs-recommendations.html.

as a ratio between free PSA and total PSA. This ratio should be greater than 25 percent which means that cancer of the prostate gland is less likely. If the ratio is less than 25 percent, then there is a stronger possibility of prostate cancer and additional testing may be indicated.

- Prostate Health Index (PHI)—This test combines total PSA, free-PSA, and pro-PSA in order to predict the presence of prostate cancer.

- PCA3—This is a urine test that detects abnormal genetic material that can shed into the urine from the prostate immediately following a prostate massage by the urologist.

- PSA velocity—A rise in PSA of greater than 0.9 mg/dL in one year—even if the level is within a normal range—increases the suspicion of prostate cancer.

- 4K Score—This blood test examines four proteins (total PSA, free PSA, intact PSA, and human kallikrein-related peptidase 2) to differentiate men at risk for aggressive prostate cancer from those with either a low risk prostate cancer or no prostate cancer at all. The 4K score is useful in helping men make a decision to proceed with a prostate biopsy. A low risk score is <7.5 percent and a high risk is >20 percent. Men with a low risk score can be advised to defer a biopsy, and they can be safely followed since they are at low risk of adverse outcomes ten to twenty years later.

There are several variations on biopsy technique. The most common method uses ultrasound to guide the biopsy needle.

A one-inch diameter ultrasound probe is gently placed in the rectum in order to visualize the prostate. A local anesthetic such as lidocaine is injected into the nerves adjacent to the prostate in order to eliminate pain during the needle biopsies. Approximately twelve biopsies are performed, during which each time you will hear a loud popping noise. But don't worry—its bark or (the noise that occurs with the biopsy device) is worse than its bite.

More recently, some prostate biopsies are performed with magnetic resonance imaging (MRI) guidance. Although the MRI is more time-consuming and expensive, MRI has the advantage of potentially localizing cancers that ultrasound can miss if the cancer is small and not in the path of the biopsy needle. Another advantage of MRI is that the aggressive tumors are more likely to be seen on an MRI than the slow-growing tumors that we may not want to treat. Currently, MRI is used for men who have undergone a normal biopsy but still have a high risk of undiscovered prostate cancer (for example, progressively rising PSA). If diagnosed with cancer, this information could also be useful for treatment planning.

Occasionally, your doctor may recommend a general anesthesia for your biopsy. For example, sometimes a larger number of biopsies are required. Also, if the biopsy is performed through the skin rather than through the rectum, using only a local anesthetic would not be very tolerable.

Before You Solve a Problem, You Must Define It

There are two terms used to characterize any type of cancer—"stage" and "grade." Your doctor will use these characteristics when it comes to planning treatment, prescribing medications, deciding on surgical treatments, or even estimating the likely long-term outcome or course of the disease.

Staging

Staging is a way to describe the extent or severity of an individual's cancer. As the tumor develops, it can invade nearby organs and tissues, or cells can break off and enter the bloodstream or lymphatic system. The cancer then spreads (becomes "metastatic") to form new tumors in other organs.

Your doctor will determine a cancer's "clinical stage" by using a combination of physical examination, x-ray imaging such as computerized tomography scan (CT scan) or magnetic resonance imaging (MRI scans), laboratory tests, biopsies, pathology reports, and even physical symptoms a patient describes such as bone and back pain. If the tumor is removed, microscopic examination by the pathologist often reveals a more complete "pathologic stage."

Prostate Cancer Staging[5]

T1: tumor present, but not detectable clinically or with imaging

T1a: tumor was incidentally found in less than 5 percent of prostate tissue resected for other reasons

T1b: tumor was incidentally found in greater than 5 percent of prostate tissue resected for other reasons

T1c: tumor was found in a needle biopsy performed due to an elevated PSA blood test

---

5. *TNM Classification of Malignant Tumours*, Eighth Edition, James D. Brierley, Mary K. Gospodarowicz, Christian Wittekind. 2017

T2: the tumor can be felt (palpated) on examination, but has not spread outside the prostate

T2a: the tumor is in half or less than half of one of the prostate gland's two lobes

T2b: the tumor is in more than half of one lobe, but not both

T2c: the tumor is in both lobes *

T3: the tumor has spread through the prostatic capsule (if it is only part-way through, it is still T2)

T3a: the tumor has spread through the capsule on one or both sides

T3b: the tumor has invaded one or both seminal vesicles

T4: the tumor has invaded other nearby structures

* It should be stressed that the designation "T2c" implies a tumor that is palpable in both lobes of the prostate. Tumors that are found to be bilateral on biopsy only but which are not palpable bilaterally should not be staged as T2c.

Although competing staging systems still exist for some types of cancer, the universally accepted staging system is that of the "TNM Classification." The "T" describes the size of the tumor and whether it has invaded nearby tissue, the "N" describes regional lymph nodes that are involved, and the "M" describes distant metastasis (spread of cancer from one body part to another). For instance, a patient with a small prostate nodule but no evidence of spread to lymph nodes or other locations would be classified as "T2aN0M0."

Prostate Cancer Grading

Cancer grading (also called tumor grading) is a system used to classify the aggressiveness of cancer cells in terms of how abnormal they look under a microscope and how quickly the tumor is likely to grow and spread. This information can be obtained from a small sample (biopsy) of the tumor. However, as in staging, the information can be more accurate by microscopic examination of the entire tumor following complete removal. Prostate cancer has a unique grading system for classifying the aggressiveness of prostate cancer tissue—the Gleason grading system.[6]

Based on how it looks under a microscope, a pathologist determines the level of aggressiveness on a scale from 1 to 5. Since more than one level can exist in the same prostate, the two most predominant Gleason "grades" are added together to give a Gleason "score" (for example, "3+4=7"). Since Gleason grades range from 1 to 5, Gleason scores will range from 2 to 10. These scores can indicate how quickly the tumor is likely to grow and spread. When more than one Gleason grade is present, the most predominant grade is written first ("4+3=7" is more aggressive than "3+4=7"). A low Gleason score means the cancer tissue is similar to normal prostate tissue, and the tumor is less likely to spread. A high Gleason score means the cancer tissue is very different from normal, and the tumor is more likely to spread.

To make things even more complicated, pathologists no longer use Gleason "grades" 1 or 2. Therefore, in essence, the Gleason System "scores" prostate cancer cells on a scale from 6 to 10. As such, the World Health Organization is replacing the

---

6. Gleason DF. Classification of prostatic carcinoma. *Cancer Chemotherapy Reports* 1966;50:125–128.

Gleason System with the new Five Group Grading System.[7] This newer system is slowly being adopted by the medical community. During this transition, both systems are usually reported in the following manner:

- Grade Group One (Gleason score ≤6)
- Grade Group Two (Gleason score 3+4=7)
- Grade Group Three (Gleason score 4+3=7)
- Grade Group Four (Gleason score 8)
- Grade Group Five (Gleason scores 9–10)

Prostate Cancer Risk Groups

Using stage, grade, and PSA levels, various methods have been proposed to categorize overall level of risk. The most widely used system was developed by the National Comprehensive Cancer Network (NCCN)[7] as follows:

- Very low risk—The tumor cannot be felt during a DRE and is not seen during imaging tests but was found during a needle biopsy (stage T1c). PSA is less than 10 ng/mL. The Gleason score is 6 or less. Cancer was found in fewer than three samples taken during a core biopsy. The cancer was found in half or less of any core.

7. Epstein JI, Zelefsky MJ, Sjoberg DD, et al. A contemporary prostate cancer grading system: A validated alternative to the Gleason score. *European Urology* (2015).

8. NCCN Guidelines for Patients, Accessed August 01, 2017. https://www.nccn.org/patients/guidelines/prostate/files/assets/common/downloads/files/prostate.pdf.

- Low risk—The tumor is classified as stage T1a, T1b, T1c, or T2a. PSA is less than 10 ng/mL. The Gleason score is 6.
- Intermediate risk—The tumor has two or more of these characteristics:
    - Classified as stage T2b or T2c
    - PSA is between 10 and 20 ng/mL
    - Gleason score of 7
- High risk—The tumor has two or more of these characteristics:
    - Classified as stage T3a
    - PSA level is higher than 20 ng/mL
    - Gleason score is between 8 and 10
- Very high risk—The tumor is classified as stage T3b or T4. The Gleason grade is 5 for the main pattern of cell growth, or more than four biopsy cores have Gleason scores between 8 and 10.

The Role of Testosterone—Fuel for the Fire?

A lot of controversy exists as to whether testosterone causes prostate cancer. Testosterone replacement can make prostate cancer grow faster and can elevate PSA levels and prostate size in those without prostate cancer. However, there is no evidence to support that testosterone actually causes prostate cancer. Most of these concerns stem from the fact that depletion of testosterone can help treat prostate cancer (discussed later in this chapter). However, this relationship does not necessarily mean bringing a man's testosterone from a deficient level to a normal level will cause a man to develop prostate cancer. After a successful treatment for prostate cancer, it is reasonable to replace testosterone with careful monitoring. After all, we do not intentionally deplete the testosterone in men who we think have been cured by their treatment. Therefore, why should we deny testosterone replacement in men suffering from the debilitating effects of a low testosterone level? Of course, this decision is best made through discussions with your physician.

Hey, You Look Good in Them Genes— the Role of Genomic Testing

Since the formation of cancer cells all starts with changes in a cell's genetic material—also known as DNA—we can perform specialized testing that can help further define the behavior of an individual's prostate cancer. So what is the difference between "genomic" testing and "genetic" testing? Genetic testing looks at your genes, whereas genomic testing looks at the genes of the cancer cells. The following tests are currently the most commonly used biomarker tests that search for specific genetic mutations that can help predict a patient's prognosis:

- Prolaris®—For initial biopsy specimens, this test can help predict the chance of dying from prostate cancer within ten years. When the prostate is removed, this test can help predict the likelihood of the PSA levels rising within ten years ("biochemical recurrence" or "BCR").
- OncotypeDX®—This is a pre-operative test of the biopsy that predicts the likelihood of higher grade or escape from the prostate upon microscopic examination of the prostate following surgical removal.

- Decipher®—This test is performed on the surgically removed prostate to predict the likelihood of metastatic spread of the disease within ten years following surgery.
- ConfirmMDx®—When a biopsy does not reveal cancer but a high level of suspicion still exists, the normal biopsy material can be tested to predict the likelihood of undiagnosed cancer elsewhere in the prostate gland.

In addition to predicting who would benefit most from aggressive therapy, biomarkers can help patients' expectations and help predict outcomes of treatment. Genomic testing can even help determine the necessity of treatment in low-grade prostate cancers (especially in older men or men with other medical conditions such as heart disease, diabetes, and hypertension).

Further testing

The PSA blood test, prostate exam, and biopsy results are the primary determinants of the extent, or "stage," of the disease. In higher risk cases, some additional testing may be recommended.

Bone scan

When prostate cancer does escape the prostate, it can travel to bone. Prior to this test, a slightly radioactive substance (radionuclide) is injected into the veins. This radionuclide then binds to prostate cancer that has migrated to the bones and is measured by a specialized camera. The radionuclide will also bind with other abnormalities of the bone such as arthritis or prior trauma. An experienced radiologist can usually differentiate these conditions from cancer.

CT Scan

Computer tomography (CT scan) is sometimes used to look for spread of prostate cancer to lymph nodes or other organs. Typically, this test is recommended for patients with a PSA level greater than twenty-five or high grade disease.

MRI Scan

MRI can also be useful for treatment planning. For instance, knowing where the cancer is located relative to the nerves responsible for erectile function can help your physician preserve this ability. We always want another thing to hang our hat on!

What's the Answer for the Cancer?

Once the prostate cancer is detected, there may be several effective treatment options. Every person's situation is unique. Factors that influence a treatment decision include age, overall health, extent (stage) of the disease, aggressiveness (grade) of the disease, and personal preference of the patient. Nine out of ten prostate cancer cases are caught early. Most of these cases are best treated with either removal or destruction of the prostate gland. However, in some men with very early and indolent disease, careful observation ("active surveillance" or "watchful waiting") is the most appropriate choice. Patients who are diagnosed late—once the disease has spread to other locations—may benefit most from hormonal or chemotherapeutic medications.

Active Surveillance

Since prostate cancer is a slow growing disease, patients with a small amount of low grade disease can often be safely observed. This option is aptly named active surveillance since it requires regular PSA blood testing and ongoing repeat prostate biopsies. PSA alone is not reliable for determining progression of the cancer. Genomic testing and MRI scan can also enhance the ability to assess who are the best candidates for active surveillance.

You may ask, "why would I consider leaving cancer untreated in my body?" In reality, active surveillance should actually be considered a treatment rather than "doing nothing." Just like the other treatment options, you should weigh all of the advantages and disadvantages. A proper candidate would have a low volume of cancer (three or less cancerous biopsy cores, each with less than 50 percent involvement) and a low Gleason score. Since prostate cancer is usually slow-growing, such a patient would have a wide window between the time of initial diagnosis and the time at which the likelihood of cure would decrease. In some men, this time span is indefinite. In others, treatment is merely delayed a few years, thereby avoiding exposure to the risks of treatment for a period of time.

If your urologist feels that you are a potential candidate for active surveillance, he may want to do some further testing. For instance, genomic testing can help determine how aggressive your cancer may be. In some patients, particularly those with an enlarged prostate, an MRI scan can search for an abnormality that may have slipped between the biopsy needles.

The consideration of active surveillance is very similar to a job search. Your biopsy and PSA results are your résumé. If your urologist then considers you to be a potential candidate, additional testing (genomic, MRI) would serve as the interview. Even if you are offered the "job" of active surveillance, you do not need to accept. In fact, at any time you can move onto other treatment prospects.

Obviously, the advantage of active surveillance is avoidance of treatment side effects. The disadvantages of this approach are the emotional toll and the remote possibility that the cancer can escape the window of opportunity for cure. In addition, those on active surveillance must commit to frequent PSA testing and repeat prostate biopsies every one to two years.

Watchful Waiting

At the risk of being caught up in semantics, watchful waiting is very similar to active surveillance, but with a much lower intention of shifting to one of the active treatments listed below. As such, repeat biopsies are seldom performed, and PSA testing is less frequent. The best candidates are men with a life-expectancy of less than ten years due to other health issues and/or age. Watchful waiting focuses mainly on prevention and early detection of the complications of prostate cancer.

Surgical Removal

The most common method of prostate removal is robotic prostatectomy. Robotic prostatectomy is a type of laparoscopic surgery—with an added layer of technology. As with other laparoscopic procedures, rather than using a large incision, the surgeon makes a button-size incision in the abdominal cavity for the insertion of a telescope. After expanding the abdominal cavity with carbon dioxide gas, several additional small incisions are made to place narrow tubes used for interchangeable instruments. Instead of the surgeon's hands directly moving the instruments, the robotic device is wheeled up to the patient, and the robotic arms are attached to the telescope and the instruments. The surgeon then sits at the control console a few feet from the patient, leaving the surgical assistant and scrub nurse at your side. One or two additional small tubes are often placed for the surgical assistant to use. The surgeon then views a highly magnified, three-dimensional image of the patient's interior structures. All movements of the camera and robotic instruments are precisely performed in real-time by the surgeon using ergonomic finger controls. The tips of these instruments can make any wrist-like turn that the surgeon so desires. The procedure is performed using instruments such as miniature tweezers and scissors the size of a fingernail (although these scissors appear to be the size of hedge clippers to the surgeon observing them on the video monitor).

Possible side effects of surgical removal include incontinence and erectile dysfunction. Your urologist can guide you on your likelihood of experiencing any of these side effects.

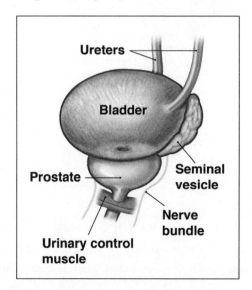

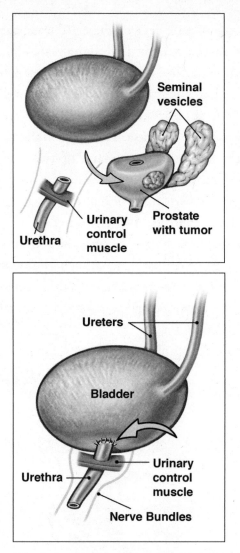

Figure 2: Prostate Removal *(C. Boyter)*

Radiation Therapy

There are many ways to deliver radiation energy to the prostate. The two most common ways are implantation of radioactive seeds—also called brachytherapy—and external beam radiation. With brachytherapy, the radioactive seeds are implanted using needles placed through the skin located between the scrotum and the anus (the area referred to as the perineum). A similar technique—high dose radiation, or HDR—involves placing high energy needles temporarily into the prostate and removing them prior to your leaving the medical facility. Usually, HDR is performed two separate times spaced one to two weeks apart.

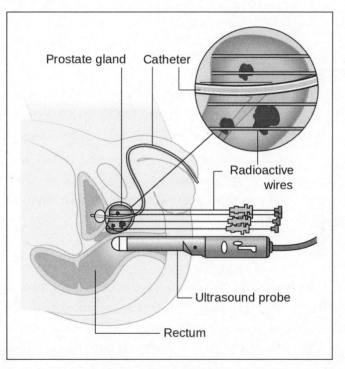

Figure 3: Prostate Seed Implantation *(Wikimedia Commons)*

Other forms of external radiation sometimes used for treating prostate cancer include proton beam therapy and stereotactic body radiation therapy (SBRT, CyberKnife®). Although these two therapies have some theoretical advantages, so far neither has been shown to be superior to the other forms of radiation with respect to cure rates or side effects.

Possible side effects of radiation can include incontinence, radiation cystitis (painful and/or frequent urination), stricture (scar tissue formation with urinary blockage), erectile dysfunction, and radiation damage to the rectum. A slight increase in subsequent (years later) bladder, rectal, and prostate (a second one) cancers has been reported. Your physician can guide you on your likelihood of experiencing any of these side effects.

Freezing

Cryotherapy is a well-established technique in which the prostate is frozen to a temperature below which no cells can live. Unfortunately, because of the high side effect profile (especially erectile dysfunction), it is seldom used for first-line therapy. However, it is still being used for focal therapy (see below) and prostate cancer recurrence following radiation therapy.

Heating

Just as with freezing, prostate cells cannot survive at high temperatures. High intensity focused ultrasound (HIFU) delivers directional and precise ultrasound waves through the rectum into the prostate, while visualizing the systematic destruction in real-time. HIFU can potentially have a more predictable destruction of prostate tissue compared to cryotherapy and radiation therapy, thereby minimizing the possibility of side effects. This technology is seeing an increase in use for focal therapy (see below).

Focal therapy

In a select group of men who have cancer in one relatively small area of the prostate, focal therapy can be used to destroy that one area, thereby directly minimizing the chance of urinary or sexual side effects. HIFU, focal laser ablation (FLA), and cryotherapy have all been used in this manner. The key to success is ensuring that the cancer is isolated to the targeted portion of the prostate. An MRI of the prostate may provide this information, but another option is to perform a saturation biopsy (three to four times the number of needles compared to a standard prostate biopsy).

Hormone Therapy

When prostate cancer escapes from the prostate to lymph nodes, bones, or other distant locations, treatment is focused on systemic therapy—medicines that treat the cancer wherever it may be lurking. Hormone therapy—more commonly referred to as androgen deprivation therapy, or ADT—has been the foundation of this treatment for many decades. The growth and survival of most prostate cancer cells are dependent on the presence of the male hormone, testosterone. In the absence of testosterone, most of the cancer cells will either die or regress. One way of achieving this goal is to remove both testicles (bilateral orchiectomy), the major source of this hormone (the adrenal glands produce one-tenth). More commonly, injections every one to six months are given to achieve the same effect. These medications include Leuprolide (Lupron®, Eligard®), Goserelin (Zoladex®), Triptorelin (Trelstar®), and Degarelix (Firmagon®). Although these drugs are widely effective at controlling metastatic prostate cancer, their effect is usually temporary. The duration of success is dependent mostly on the extent and aggressiveness

(Gleason score, rate of PSA rise) at the time of initiating this therapy. When these drugs fail, drugs such as Abiraterone (Zytiga®) and Enzalutamide (Xtandi®) can further block the stimulatory effect of male hormones on prostate growth.

Side effects of ADT include hot flashes, decreased sexual desire, erectile dysfunction, fatigue, depression, breast tenderness/growth, and osteoporosis. Some studies have shown that ADT can raise the risk of a cardiovascular event, but this risk is small—especially when compared to the risk of cancer progression.

ADT is sometimes used in patients undergoing radiation therapy for cancer isolated to the prostate region, either as an initial therapy following diagnosis or following surgery. The theory behind this approach is that the medication will make the prostate cancer cells more susceptible to the radiation.

Hormone Resistance

When ADT is no longer effective at controlling the growth of prostate cancer, the condition is referred to as castrate-resistant prostate cancer, or CRPC. These unfortunate patients have a rising PSA and/or growth of tumors on x-ray, despite maintaining a low testosterone level.

Immunotherapy

The only commercially available immunotherapy for prostate cancer is Provenge® (sipuleucel-T). This therapy is also called a therapeutic vaccine. Like a preventative vaccination, it will cause your immune system to fight the disease—in this case, the prostate cancer.

The treatment involves donating blood that is then filtered by a special lab to isolate specialized cancer-fighting white blood cells. These cells are stimulated with proprietary proteins and then reinfused into your bloodstream. This process is repeated every two weeks for a total of three treatments. Interestingly, although this treatment has been shown to prolong life, the PSA values often do not decrease.

Chemotherapy

Chemotherapy is usually used in CRPC patients who either have symptoms from their cancer or when all other treatments have failed. The two most common chemotherapy drugs are Docetaxel (Taxotere®) and Cabazitaxel (Jevtana®). Possible side effects include hair loss, mouth sores, nausea, vomiting, diarrhea, and fatigue.

Bone Health in the Patient with Advanced Prostate Cancer

One of the major concerns with androgen deprivation therapy is loss of bone density leading to osteoporosis (or its lesser counterpart, osteopenia). In older men who may already be experiencing some bone loss, a bone density scan (not to be confused with a bone scan mentioned earlier) should be performed to measure a baseline prior to initiating ADT.

Lifestyle management is the foundation to prevention of bone complications in prostate cancer patients receiving ADT. These men should be advised to avoid smoking and excessive alcohol consumption. They should also perform regular weight-bearing aerobic and resistance exercise to stimulate bone growth and to increase muscle strength to support their skeleton. To prevent calcium and vitamin D deficiencies, they should take daily supplements of at least 1200 mg of calcium and 800 IU of vitamin D.

All men receiving ADT should have their bone density monitored at least yearly. If their bone density decreases despite the above measures, they should follow the routine guidelines for the management of osteoporosis (see Chapter 9). In men with castrate resistant prostate cancer and bone metastases (spread to the bone), pharmaceutical intervention is mandatory since fractures are more common in this scenario. Commonly used medications include zoledronic acid (Zometa®) and denosumab (Xgeva®, Prolia®). Your physician can determine which regimen is right for you.

In men with CRPC and symptomatic (painful) bone metastases, Xofigo (radium Ra 223 dichloride) injection is given to prevent fracture and reduce the pain. This intravenous drug binds to the affected bone and delivers radiation directly to the cancerous lesions. Chemotherapy should not be given at the same time since together they can suppress the bone marrow.

Managing the Side Effects of Treatment

The best physicians are the ones who are there for you when things are not going just right. Most men who experience incontinence following surgery or radiation will recover with time and/or pelvic floor exercises (Kegels). For the remaining minority, there are several surgical solutions, all of which are discussed in Chapter 11. Unfortunately, medications seldom work in these scenarios.

Recovery of erectile function is dependent on three things—pre-treatment function, the nature of the treatment, and post-treatment rehabilitation. The more problems a man has with his erections prior to treatment, the more likely he is to lose that function. Of course, more extensive disease requires more extensive treatment which can lead to damage of the neighboring nerves that supply erectile function. In any case, encouraging blood flow as soon and as often as possible following treatment will maximize the chances of recovery. Temporary lack of blood flow to the penis can lead to permanent scar tissue formation in the microscopic vascular spaces within the penis. Aggressive rehabilitation can help prevent this process. Erectile rehab can include any combination of the following:

- Daily dosing of medications such as sildenafil (Viagra®), tadalafil (Cialis®), and vardenafil (Levitra®, Staxyn®)
- Daily use of a vacuum erection device
- Regular use of prostaglandin PGE penile injections (Caverject®, Edex®) or urethral suppositories (MUSE®)

No better example of "use it or lose it" exists! Management of erectile dysfunction following prostate cancer treatment is discussed in detail in Chapter 4.

The side effects unique to radiation therapy can usually be managed with oral medications for urinary symptoms and topical creams for rectal symptoms.

Choosing a Treatment

Many factors go into deciding the best course of action for treating prostate cancer. These include pre-existing symptoms, extent/volume of disease, Gleason grade, PSA level, age, and emotional factors. Every treatment choice carries some risk, even if performed with minimal or no incisions. Whatever decision you chose, extensive experience of the treating physician is essential in order to maximize a good outcome. Of course, there is no substitute for a direct conversation with your doctor. In the end, comfort with your decision is as important as the decision itself.

Usually the first step in determining your best treatment option is considering whether you are a candidate for active surveillance. For those who conclude that they are best suited for active treatment, most will narrow the choices down to surgical removal or radiation. Choosing between these two treatments can be daunting.

The biggest advantage to radiation therapy is that it is easier to undergo than surgery. Even with robotic technology, its small incisions are not as small as the openings made by a couple of dozen needles inserted into the skin behind the scrotum when radiation seeds are implanted. General recovery from robotic prostatectomy is usually two to three weeks, whereas recovery from radioactive seed implantation is one to two days. Many patients who choose the radiation route will also undergo daily external beam treatments for four to seven weeks.

The biggest advantage to surgical removal is the information learned that is not available through other treatment methods. Once the prostate is removed, it can be fully analyzed to determine the extent, location, and grade of the disease within the prostate and seminal vesicles (and lymph nodes if necessary). More important, the ability to monitor a patient for possible recurrence is dramatically enhanced. When the prostate is removed, the PSA blood test should become undetectable (<0.1) within six weeks if all the cancer cells have been successfully eliminated. Following radiation, the PSA in some cases may never become undetectable since the prostate is still present in the body. Furthermore, even with a fully successful cure, the PSA can dramatically fluctuate for the first several years following the radiation treatment. Although most properly selected patients can be cured with either surgery or radiation therapy, detecting the minority of treatment failures in a timely fashion is critical to selecting the appropriate back-up option. With PSA as the only guide, it can be difficult to distinguish between the expected PSA changes—or "bounce"—following radiation from surviving prostate cancer cells lurking in the shadows.

The most significant disadvantage to removal of the prostate is a higher possibility of long-lasting bladder control problems as compared to radiation therapy. The most bothersome possible disadvantage to radiation therapy involves difficulty emptying the bladder. In severe cases, patients can experience debilitating frequent urination (including multiple disruptions of sleep), decreased urinary flow, and pain with urination. Furthermore, in those patients who already experience frequent urination (day or night) or decreased urinary flow, the likelihood of long-lasting symptoms following radiation therapy increases dramatically. In fact, patients with these pre-existing symptoms would most likely notice an improvement after prostate removal.

So, which of these two treatments is better at curing prostate cancer? To determine this, we would need to conduct a very large study of patients (many thousands) willing to have their treatment chosen by the flip of a coin. Since that is not possible, several theories have been proposed. With radiation, since the areas around the prostate also receive a dose, cancer cells that have just penetrated through the capsule can potentially be eliminated. However, this theory is difficult to prove since we do not know which patients have this scenario (no examination of the prostate by the pathologist). However, radiation can be given following surgery for those patients proven to have disease just outside of the prostate, thereby avoiding radiation to the surrounding areas in most patients.

Sometimes age can be a consideration in the treatment decision. By no means is the life of a forty-year-old more valuable than a seventy-year-old. However, since prostate cancer is usually slow-growing, treatment failures can often be managed in older patients. In the younger patient, surgical removal provides an improved knowledge of disease status with potentially better and more timely backup options. Whereas radiation can easily be given after surgery, rarely is surgical removal possible following radiation. In addition, younger patients are much more likely to avoid some of the side effects associated with prostate removal. Since some of the risks of radiation can be delayed by years, these issues can be more significant for younger patients. For instance, the risk of secondary cancers of the rectum, bladder, or prostate (a new prostate cancer) can increase many years after the radiation treatment. Of course, any of the potential side effects to the bladder or rectum from radiation therapy are avoided with surgical removal.

Questions You Might Want to Ask Your Doctor

- What is my cancer stage and grade?
- Do I need any other tests?
- Am I a candidate for active surveillance?
- What are *my* chances of cure in *your* hands?
- What are *my* chances of urinary, sexual, or other side effects in *your* hands?
- What are all my treatment options?
- What do you consider the best treatment for my cancer?

- How many procedures have you performed on patients like me?
- How often do you perform these procedures and how many have you performed in total?
- Do *you* perform the entire procedure or treatment?
- What are my options should my initial choice prove unsuccessful?
- How often will I need to have follow-up exams, blood tests, and imaging tests?

So Whatever Happened to Jerry?

Based on Jerry's biopsy, PSA, and genomic testing, Dr. Samples determined that his prostate cancer was intermediate risk. Therefore, he was not a good candidate for active surveillance. Since Jerry did not have any symptoms, he would not have known this life-saving information had he not underwent PSA blood testing. After many discussions with several physicians, he and his wife decided that robotic prostate removal was his best option. Several months after the surgery, his fear of side effects subsided when his performance in both the bathroom and the bedroom returned to the way it used to be.

Bottom Line

Prostate cancer is the third leading cause of cancer death in American men, behind lung cancer and colorectal cancer.[9] Starting at age forty, men should learn their risk of developing

9. "Cancer Facts & Figures 2017." American Cancer Society. Accessed September 01, 2017. https://www.cancer.org/research/cancer-facts -statistics/all-cancer-facts-figures/cancer-facts-figures-2017.html.

a harmful prostate cancer. It all starts with information easily obtained by asking the right questions and having meaningful discussions with their physician. Only then can a man make an informed decision of how they should be screened and/or treated for prostate cancer.

CHAPTER 4

Erectile Dysfunction/ Impotence—When "Mr. Happy" Goes to Sleep or Developing a Weapon of Mass Destruction

"Man survives earthquakes and all the tortures of the soul. But the most tormenting tragedy of all times is the tragedy of the bedroom."—Tolstoy

Mr. Johnson, age fifty-six, has been experiencing progressive difficulty obtaining and maintaining an erection adequate for sexual intimacy. He stated succinctly that his "Johnson" wasn't working. He has a history of hypertension, elevated cholesterol levels, and adult onset diabetes. He is also fifty to seventy-five pounds overweight. He is experiencing mild depression about the erectile dysfunction and has noted that his relationship with his wife of thirty years has deteriorated.

Of all the chapters in this book, it is this chapter that will probably impact every male reader, as impotence or erectile dysfunction (ED) affects most men at some time during their lives. Just as Leo Tolstoy said in 1920, a tragedy in the bedroom is often devastating to a man and his partner. It impacts a man's self-confidence, his masculinity, his productivity in the work place, and nearly every aspect of his life. This chapter is going to explore the topic in details, put to rest many of the myths associated with ED, and discuss the treatment options that are available so that a man no longer needs to suffer the "tragedy of the bedroom."

You Are Not Alone

Erectile dysfunction (ED) is a significant problem. There are over 150 million men in the US, and nearly thirty-three million of those men experience a failure in the bedroom. The Massachusetts Male Aging Study was probably the first study that brought to light how common erectile dysfunction is. This study demonstrated that fifty-two percent of all men between the ages of forty and seventy have some degree of erectile dysfunction.[1]

Although the disease is considered a benign disorder, it may have a dramatic impact on the quality of life of many men, as well as that of their sexual partners. ED often results in anxiety, depression, and lack of self-esteem and self-confidence, all of which can perpetuate the disorder.

---

1. Feldman, H.A., Goldstein, I., Hatzichristou, D.G., Krane, R.J. "Impotence and it's Medical and Psychosocial Correlates: Results of the Massachusetts Male Aging Study." *Journal of Urology.* 151, no. 1 (January 1994): 54–61.

Brief History of Treating ED[2]

Sex and intercourse is a basic human need, common to all people from the beginning of recorded history until this very day. Erectile dysfunction is by no means a recent development in medical history, nor is it one that's been kept quiet or away from public discussion. The failure of men to rise to the occasion has been a of interest to men, and also to women, since the dawn of human history.

Ancient Egyptians had a rich and varied sexual life that is chronicled in numerous hieroglyphs and papyri. In the times of the pharaohs, the Egyptians defined impotence and recorded methods of treating this malady.

Egyptian physicians used magical spells and incantations to improve the sexual prowess of their patients. Perhaps this was the first written evidence of treating erectile dysfunction of psychogenic origin.

Hippocrates, the father of modern medicine, wrote in one of his texts, *The Aphorisms*, ". . . when two, three, or even more attempts are attended with no better success, they think they have sinned against the God and attribute thereto the cause."

Means of stimulating sexual desire in men was addressed in Ancient Greece and Rome. An example is with Pliny the Elder's *Natural History* advocating various root vegetables and plants as potent aphrodisiacs.

The ancient Chinese correctly correlated aging with ED: "At the age of forty-eight, the potency of a male is reduced or even exhausted. As his genital secretions are exhausted, he can

2. Schultheiss D, Musitelli S, Stief CG. *Classical Writings on Erectile Dysfunction*. Berlin: ABW Wissenschaftsverlag GmbH, 2005.

no longer hold it erect and, at this point, he cannot generate any more."

Ancient India also had solutions for ED. The Kama Sutra is the earliest example of a love-making manual in the history of the world. They had potions consisting of the cooking the testicles of a goat in milk, which were purported "to make a man as powerful as a bull."

Of course, there were advances from these ancient times to the Renaissance where the famous drawings of the anatomical structures in the male and female pelvis were demonstrated to the medical community.

However, one of the greatest discoveries was the use of phosphodiesterase-five inhibitors to increase blood supply to the penis. This resulted in the development of sildenafil or Viagra, which has made it possible for millions of men worldwide to solve their problem of ED.

No book on ED would be complete without mention of Alfred Kinsey and Masters and Johnsons. They conducted groundbreaking research on human sexuality and clearly made the distinction between psychological and physical impotence.

All told, there have been thousands of studies and reports on treatments for ED made over the last two millennia. Most of these are based on the discoveries of innovative and creative physicians and scientists who have made it possible for millions of men to be able to successfully engage in sexual intimacy.

Definition of ED

Erectile dysfunction is persistent and consistent failure to achieve and maintain an erection adequate for the sexual needs of a man and his partner. It is a common condition affecting thirty-three million American men. An occasional failure to achieve an erection happens to most men at some time during their lives. However, when it happens most of the time, it becomes a problem that affects a man's masculinity, erodes his confidence, and even impacts his performance and productivity in the workplace. The good news is that nearly every man with ED can be successfully treated for this common medical problem.

How Does an Erection Occur?

An erection is similar to filling a tire using an air compressor. In order to fill a tire, you need to have the air compressor plugged into the wall and have an electric current flow to the air compressor. The air compressor has to be turned on and has to take air from the environment and make it flow through the tube from the air compressor to the tire. The tubing has to be open or patent without obstruction or kinks which could impede the airflow. The tube has to be attached to the nipple on the tire, and the tire must be a closed system without any holes that could allow the air to escape.

The same principles of filling a tire with air hold true for a man's erection. The erection starts in the brain and sends a message via nerves in the spinal cord to open the blood vessels that allow blood to flow into the penis. The heart has to be working properly to pump the blood into the blood vessels that supply the penis. When more blood flows into the penis through the arteries than leaves the penis through the veins, an erection will occur.

Failure of any component from the brain to the end organ, the penis, will result in erectile dysfunction.

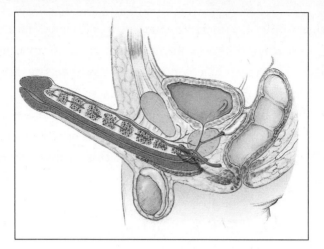

Figure 1: Erect Penis *(Coloplast)*

Causes of ED

ED can broadly be divided into two categories: psychological and physical. In the past, most ED was thought be psychological in nature. Now we have identified many physical causes such as injury to nerves and conditions affecting the blood vessels as the most common causes of ED. For men over the age of fifty years, physical causes are the most common. In younger men, especially under the age of forty years, psychological causes prevail.

Of the physical causes, vascular disease is the most common. Arteriosclerosis, or hardening and narrowing of the blood vessels that supply the penis, causes more than fifty percent of the cases of men with ED. Increased levels of cholesterol and hypertension or high blood pressure can also narrow the blood vessels and decrease the blood flow into the penis causing ED. Diabetes affects the nerves and the blood vessels resulting in ED in nearly fifty percent of men with diabetes. Neurologic causes, which impair the nerves to the penis or affect the brain and spinal cord, include spinal cord injury, multiple sclerosis, Parkinson's disease, stroke, and Alzheimer's disease. Significant pelvic trauma (such as a fracture of the pelvis), pelvic surgery (such as removal of the entire prostate gland for prostate cancer), and bladder or colon surgery can result in ED. Radiation therapy to the prostate gland or bladder can result in injuries to the nerves and blood supply, which can impact a man's ability to have an erection. Peyronie's disease is a condition that results in the formation of thick scar tissue in the penis that can result in curvature of the penis and can alter the blood supply, thereby resulting in ED.

Medications are also a culprit. There are hundreds of medications such as blood pressure medications, heart medications, antidepressants, tranquilizers, pain medications, and sedatives, just to name a few. Table 1 is a list of some of the most common culprits or medications that have side effects of ED. The male hormone testosterone, produced in the testicles, is primarily responsible for libido (or sex drive) and deficiencies of testosterone can affect a man's erection. Venous leak is a condition that results in blood leaving the penis as fast as it enters and is usually associated with failure to maintain the erection once it has been achieved. It would be similar to trying to fill the tire with air from the air compressor while there is a hole in the tire. As fast as air enters the tire, it leaves through the hole and the tire remains flat. This is the same problem of venous leak in a man with ED, where blood enters the penis but leaves just as quickly through the abnormal veins.

Miscellaneous causes include smoking (which affects the blood vessels), use of illicit drugs such as cocaine and excessive alcohol consumption (Shakespeare made mention of this in *Macbeth*: "alcohol provokes, and unprovokes; it provokes the desire, but it takes away the performance . . ."). Chronic

Table I
Drugs that cause ED

Antihypertensives (Clonidine, Reserpine, Beta-blockers, Verapamil, Methydopa)

Diuretics (Thiazides and Spironolactone)

Cardiac/Circulatory (Clofibrate, Gemfibrozil, Digoxin)

Tranquilizers (Phenothiazines, Butyrophenones)

Anticholinergics (Disopyramide and Anticonvulsants)
Antidepressants (Tricyclic antidepressants, AAOIs, Lithium, and SSRIs)

Hormones (Estrogen, Corticosteroids, Cyproterone acetate, 5-Alph a reductase inhibitors, LHRH agonists and antagonists)

H2 antagonists (Cimetidine and Ranitidine)

Cytoxic agents (Cyclophosphamide, Methotraxate, Roferon-A)

Table 1: Medications with side effects of erectile dysfunction.

or long-standing liver and kidney disease can impact a man's erection. Other hormonal causes include too much or too little thyroid hormone and an elevation of prolactin levels, which can decrease the testosterone level. Finally, obesity is also associated with increased incidence of ED, possibly because of the association of obesity with diabetes, hypertension, and high cholesterol levels which are all co-morbid conditions associated with ED.

It is important to mention marijuana as a potential cause of ED since it is widely available for both medicinal and recreational purposes. There are cannabinoid receptors are present in the penile tissue and so it is possible for the tetrahydrocannabinol, the active ingredient in marijuana, to impair penile function. In some men, this can lead to ED.

Marijuana can cause euphoria followed by drowsiness and slowed reaction time. Some men may have diminished desire for sex after smoking marijuana. Also, marijuana increases the levels of dopamine in the body. Dopamine is a hormone that can affect mood and feeling. If a man gets used to these artificially high levels of dopamine, then he may find that his body's natural level of dopamine may not be enough to stimulate him sexually. Marijuana can also affect the circulatory system and cause blood pressure and heart rate to increase. Both of these medical conditions are risks for ED.[3]

Another concern for men that may cause their ED is excessive or improper bicycle riding. When you're riding a bicycle, your weight is being focused on the perineum, the area between the rectum and the scrotum, and that's where the arteries and nerves that feed the penis are located. Since the arteries are essentially unprotected, they're prone to damage from constant pressure from the bike seat. When a man sits on a bicycle seat, he's putting his entire body weight on the artery that supplies the penis. When the artery is compressed, there is a decrease in the necessary oxygen the tissue needs in order to properly function.

Prolonged or excessive bike riding can injure the blood supply to the penis, making erections difficult or impossible. You can anticipate injury if after riding you experience numbness in the genital area. The longer that the numbness lasts is usually a barometer of how much injury you are creating for those important blood vessels that supply the penis.

---

3. Wu B, "Marijuana and erectile dysfunction: What is the connection?" Medical News Today, https://www.medicalnewstoday.com/articles /317104.php.

Psychological causes, which only cause ED in ten percent of men who suffer from ED, can occur with depression and recent emotional trauma, such as loss of a partner, divorce, stress in the work place, or job loss. Psychological causes can also occur with deterioration in the relationship between a man and his partner or if there are problems and distractions with other family members. A common situation in young men is referred to as "performance anxiety." This situation occurs when a young man is anxious about his ability to perform, especially with a new partner. The ED in this scenario becomes a cyclical event as the erectile failure itself creates more anxiety on each subsequent desire to be sexually intimate.

Priapism: An Erection That Will Not Quit or When Hard is Horrible!

We know it sounds like a dream come true when you hear about the erection that lasts for four hours after taking one of the oral medications for ED. Nothing could be farther from the truth. This situation, priapism, is a medical emergency and needs immediate medical attention. If you don't get treatment right away, it can lead to permanent ED.

Excessive blood flow to the penis or blood that is trapped in the penis can cause a very painful erection that lasts for more than four hours.

There Are Two Main Types of Priapism

Low-flow or ischemic priapism happens when blood gets trapped in the erection chambers. Most of the time, there's no clear cause, but it may affect men with blood disorders such as sickle-cell disease or leukemia.

High-flow or non-ischemic priapism often happens when there is an injury to the penis or the area between the scrotum and anus, called the perineum, and there is a rupture of an artery that supplies the erection chambers or corporal bodies of the penis.

The most common cause of priapism occurs following the use of injections of medication, such as trimix, prostaglandin, phentolamine, or papaverine into the penis (see self-injection therapy, Page 84). Other drugs that may cause priapism include some of the drugs used to treat depression and mental illnesses such as trazodone (Desyrel), or chlorpromazine (Thorazine). Also, priapism can occur following the use of street drugs like heroin and cocaine.

The goal of treatment for priapism is to make the erection go away in order to prevent permanent ED. Treatment may consist of removing the blood trapped in the penis. For low-flow priapism, your doctor can inject drugs called alpha-agonists into your penis to reduce the blood flow to the penis. This may reduce the pain and swelling. Another treatment option that has been successful is terbutaline—five milligrams, taken orally. The take home message is that an erection lasting four hours is not a dream come true, but nightmare that needs immediate medical attention.

Pop Goes the Weasel: Penis Fracture

We know that most fractures involve bones in the arms and lower extremities. However, there are other fractures that can be just as devastating and that includes fracturing the penis, even though there are no bones made of calcium in the penis. (The only mammal that has a bone in the penis is the whale!)

Penile fractures happen only when you have an erection. When you're soft, the pressure inside your penis is low, so it's more able to bend and withstand unexpected forces. However, when the penis is erect and hits a hard structure such as the woman's pubic bone, the penis cannot bend easily, leading to huge increases of pressure in the penis and a blow out or fracture.

In most cases there is a loud popping sound when the fracture occurs and is usually heard by both partners. This is followed by intense pain and then swelling of the penis.

A fracture of the penis is a medical emergency and requires immediate attention and surgery to repair the fracture.

Take home message: ED is a common problem affecting millions of American men. You are not alone. Although ED is more common in middle age and older men, it does not need to be a consequence of aging. Older men who are healthy, have an available partner, and are interested in engaging in sexual intimacy can be helped and can successfully engage in intimacy.

How is ED diagnosed?

Like most medical conditions, a careful history is taken, a physical exam conducted, and laboratory testing is ordered.

Your doctor will begin by taking a careful history. He may ask: How long have you had ED? What other diseases or conditions do you have? What medications, including over-the-counter medications or supplements, do you take? What treatments have you tried in the past? What is the impact of ED on your partner?

Next is a physical examination, which includes taking your blood pressure, an examination of the penis and testicles, and a rectal exam to evaluate your prostate gland. The physical examination also includes an evaluation of the nerve supply to the lower extremities and an evaluation of the blood vessels in the legs and feet.

You will probably be sent for a urinalysis and a few blood tests, which include a testosterone level, cholesterol level, blood sugar, and liver, kidney and thyroid function tests. If the testosterone level is low, a prolactin level and other blood tests associated with testosterone secretion may be obtained. The urine examination is for protein and glucose (sugar), which can indicate kidney disease or diabetes, respectively.

Additional tests may also include a penile color duplex ultrasound, which is used to evaluate the blood flow to the penis. The duplex Doppler study is usually accompanied by an injection of a drug or drugs into the penis in order to stimulate the blood flow to the penis. Normally, these injections create an erection that lasts about twenty minutes and allows the doctor to see if the blood vessels respond appropriately to the medication.

A nocturnal penile tumescence test is only occasionally ordered to differentiate psychological causes from physical causes. This test is based on the fact that normal men have three to five involuntary nighttime erections that can be recorded with a band placed loosely around the penis and attached to a monitoring device. If a man obtains an erection, there will be a subsequent increase in girth of the penis, which is recorded on a computer monitor. The presence of normal nighttime erections indicates that the nerve and blood supply to the penis are intact and that the problem is probably related to psychological causes rather than a physical cause.

Another option for the nocturnal tumescence test can be easily accomplished by using a few US postage stamps placed around the circumference at the base of the penis before a man goes to sleep. If a man has a normal nocturnal erection, there will be an increase in the girth of the penis, which will break the perforations between the stamps indicating that the man has a good erection and that he should be able to make penetration at the time of sexual intimacy.

How Is ED Treated?

First, the good news: Nearly everyone with ED can be helped. There are many treatment options available to all men who suffer from ED. You just have to find one that meets your situation and your pocket book. The best patients are those who are aware of the options and then, with the help of their doctors, select treatments that work best for them. Most men start with the least invasive treatment options and then proceed to additional treatments if the less invasive treatments are not effective.

Perhaps the easiest and least expensive treatment is lifestyle modification. Men who are smokers will find that there will be improvement in their erections when they stop smoking cigarettes. The active ingredient in cigarettes, nicotine, results in a decrease in the blood flow to the penis. Prolonged smoking over many years may permanently narrow the blood supply to penis, which will preclude the penis from becoming rigid enough for vaginal penetration.

Men who are overweight will notice an improvement in their ability to engage in sexual intimacy by losing weight and beginning a physical exercise program. For men with high cholesterol levels, dietary modification may decrease the cholesterol levels and improve the blood supply to the penis resulting in better erections. A question we often ask our overweight patients is the following: Mr. Smith, would you like to take a pill that improves your energy level, increases your sex drive, decreases your blood pressure, decreases your cholesterol level, decreases your risk of prostate and colon cancer, enhances your muscle mass, improves your mood, lifts your depression, and makes your penis one to one-and-a-half inches longer? Every man says, "Yes, I'd like that pill." We respond, "Mr. Smith, it isn't a pill. It's exercise!" Yes, exercise, improvement in nutrition, and weight loss will do all of the above, and the loss of the abdominal protuberance will give the appearance that the penis is actually longer!

Excessive use of alcohol also can negatively impact a man's erection, and decreasing the alcohol consumption will significantly improve a man's sexual function. Alcohol in small amounts, one to two drinks per day, may serve as a social lubricant and a sexual facilitator. Excessive alcohol may result in difficulty obtaining and holding an erection adequate for sexual intimacy. Also, it is important to mention that excessive alcohol can result in delayed orgasm or inability to have an orgasm, which can be a source of frustration for both the man and his partner. Excessive use of alcohol over many years can result in permanent liver damage. This can create hormone imbalance with an overproduction of the female hormone, estrogen, and a decrease in testosterone production. Both situations lead to a decrease in libido and sexual performance.

Men who use medications with the side effects of ED or decreased sex drive can speak to their doctor about the adverse side effects. The doctor may then decrease the dosage of the medication or change to another class of drugs that do not have the adverse side effect of ED or a decrease in sex drive or interest. For example, there are anti-depressants that have the side effect of ED, which can be changed to another medication, such as Wellbutrin, which is much less likely to cause ED and can still be effective for treating depression.

For men who are experiencing ED associated with bicycle riding, we recommend the following:

- Penile numbness and excessive genital shrinkage are warning signs that there may be too much pressure on your perineum. The nerves in the perineum are being pinched, which means the artery that feeds the penis is also being compressed.

- Make the following changes in your riding style and/or your positioning on the bike: 1) Make sure your saddle is level, or point the nose a few degrees downward. 2) Check to see that your legs are not fully extended at the bottom of the pedal stroke. Your knees should be slightly bent to support more of your weight. 3) Stand up every ten minutes or so to encourage blood flow to the penis.

- There are a multitude of anatomic racing saddles on the market, ranging from ones with a flexible nose to models with a hole in the middle. You may want to experiment with a wider, more heavily padded brand or a "double bun seat" that places the weight on the bones and off of the perineum.

- Heavier riders may be more at risk of arterial compression damage because of the greater weight that's placed on the perineum, the area between the scrotum and the rectum. If you're in this category, you should consider a wider saddle with extra padding.

- When riding a stationary bike, the tendency is to stay seated and grind against big gears for long periods. Get off of the seat as frequently as you would on your regular bike and be certain that it's set up the same in regards to riding position.
- Get off of the seat when riding over rough or irregular terrain. Use your legs as shock absorbers.

Our take home message on bicycle riding: Most men are not aware of the relationship between their bike and their erections. Our final advice for good health is that men shouldn't necessarily ride farther but ride a lot smarter.

If life-style modifications are ineffective, treatments can be organized into first line, second line, and third line treatments.

First Line Treatments—Pills for the Penis

First line treatments begin with oral medications. These medications are the phosphodiesterase-five inhibitors, which consist of sildenafil (Viagra), tadalafil (Cialis), Staxyn, and vardenafil (Levitra). The FDA first approved Viagra in 1998. These drugs are indicated in men who have ED and no contraindications to their use. The main contraindication is concurrent use of nitrate therapy for men with cardiovascular disease. This also includes men who have heart disease and carry a small container of nitrates (sublingual nitroglycerin). Phosphodiesterase-five inhibitors dilate the blood vessels to the penis but they also dilate blood vessels elsewhere in the body. Nitroglycerin also dilates the blood vessels. Men who are using the ED drugs and then take a nitroglycerin tablet may have excessive dilation of the blood vessels in the body at the expense of the blood supply through the coronary arteries to the heart. If a man is having chest pain as a result of poor blood flow to the heart, the combination of both medications will actually rob the heart of the oxygen necessary for normal functioning of the heart. It would be like using one air compressor to fill all four tires on a car at the same time. It can't be done. Remember the movie, *Something's Got to Give*? Harry Sanborn, played by Jack Nicholson, took nitroglycerin after using Viagra and was rushed to the emergency room. Using both phosphodiesterase-five inhibitors and nitroglycerin can result in a dangerous decrease in the blood supply to the heart.

There is also a precaution about using any of the oral medications in conjunction with alpha-blockers such as Rapaflo, Flomax, or Uroxatral, the medications that are often used to treat an enlarged prostate (see Chapter 2). For that reason, we recommend for men who use alpha-blockers, that they don't take the medication at the same time as using the ED oral medications. For example, if a man regularly has sexual intimacy in the morning, then he should use the ED drug in the morning and the alpha-blocker in the evening.

We don't have a preference of one first-line drug over another, as they all have the same mechanism of action. Viagra lasts four to eight hours after ingestion. The drug should be taken on as much of an empty stomach as possible as a full stomach will delay its absorption and take longer for it to become effective. Cialis lasts thirty to thirty-six hours after ingestion and Levitra last four to eight hours, but appears to have a shorter time period to start working. All three drugs require genital stimulation in order for a man to achieve an erection. Therefore, a man who plans to have sexual intimacy late at night may take the drug in the early evening before going out for dinner and will be ready to successfully engage in sexual intimacy. Table 2 is a summary of the three most commonly used oral drugs used for treating ED, the dosage, and side effects.

Table 2 Dosage and side effects of Phosphodiesterase-five inhibitors		
Viagra (sildenafil)	25, 50, 100mg	headaches, flushing, visual disturbances, nasal congestion, GERD (gastroesophogeal reflux)
Levitra (vardenafil)	10, 20mg	headache, flushing
Cialis (tadalafil)	5, 10, 20mg	headache, back pain, muscle cramps

It has been our experience that it is common for one of the phosphodiesterase-five inhibitors to be effective for several months or even years, and then cease to produce an erection adequate for sexual intimacy. We have found changing to one of the other phosphodiesterase-five inhibitors may have a beneficial effect.

Second-Line Treatment

When the phosphodiesterase-five inhibitors are no longer effective, the next treatment options consist of a vacuum device, injection therapy, or suppositories inserted into the end of the penis.

When Drugs Don't Work—Make Room for the Vacuum

A vacuum erection device is a mechanical, nonsurgical method of filling the penis with blood, thus creating an erection that is hard enough to achieve penetration. It is based on the principle that placing the penis in a vacuum chamber or plastic cylinder can produce an erection. Air is then removed from the cylinder by either a manual or electric pump creating a vacuum or negative pressure around the penis. When the penis becomes engorged with blood, a tight elastic or rubber band is released from the plastic cylinder onto the base of the penis thus trapping the additional blood in the penis. At this time, the man can begin to engage in sexual intimacy.

We recommend that the elastic or rubber band be left on the penis for no longer than thirty minutes. Because of the constricting band at the base of the penis, most of the men are not able to ejaculate, as the semen is trapped behind the constricting band. Although no fluid exits the penis, the constriction band will not prevent the sensation of climax. The side effects include mild pain when first used, transient bruising of the skin on the penis, and numbness of the penis that is relieved when the constricting band is removed.

Vacuum devices are often recommended for men after prostate cancer surgery where the entire prostate gland is removed. This allows for blood to artificially be brought into the penis after surgery and is believed to reduce shortening of the penis that often occurs after surgical removal of the entire prostate gland.

The Shot That Hits the Spot

Self-injection therapy is also an option when the oral medications are not effective. This is an effective method of producing an erection firm enough for penetration by injecting a small amount of medication directly into the shaft of the penis and the underlying corporal bodies (erection chambers). The medication is often referred to as trimix, which is a combination of alprostadil, papaverine, and phentolamine; however, an injection of only prostaglandin is available. A small amount of this medication is injected directly into the corporal bodies fifteen to twenty minutes before engaging in sexual intimacy using a very tiny needle that causes minimal discomfort. The proper dosage of trimix is determined using test injections under doctor supervision in the office setting. An erection will usually occur within ten minutes after the injection and the erection will be nearly indistinguishable from a natural erection. An erection will usually last twenty to thirty minutes. Efficacy rates for trimix of approximately seventy percent have been reported, and it is often effective for those men who do not respond to oral drug treatment.

The side effects usually consist of mild penile pain in about twenty-five percent of men, usually at the injection site. This usually subsides in a day or two. Most men do not complain of the pain and only a few men stop using these injections because of this side effect.

We advise men to limit the use of injections to two to three times per week and never more than once every twenty-four hours. We also recommend that the man use different injection sites on the penis. If the trimix is injected at the same site, there is a risk of scar formation and the development of Peyronie's disease.

The most significant problem with self-injections of trimix is the development of priapism. Priapism is the occurrence of a prolonged erection in the absence of sexual excitement. It's an erection that will not go away and come again some other day! Priapism lasting more than four hours is a medical emergency and the man with a sustained or prolonged erection must go immediately to a doctor or the emergency room. There are medications, phenylephrine or adrenaline (epinephrine), which can be injected into the penis that will reverse the priapism and allow the penis to become soft or flaccid once again. If the antidote injections for priapism are unsuccessful, urgent surgical referral is required, as surgery may occasionally be necessary to reduce the priapism.

Pellets for the Penis

Another method involves inserting a tiny drug-carrying pellet, called MUSE (Medicated Urethral System for Erection), into the opening of the urethra, the tube that transports urine through the penis from the bladder to the outside of the body. This opening is commonly found on the head of penis. This pellet, about the size of a grain of rice, is inserted into the urethra, and the drug is absorbed into the surrounding tissues of the penis to produce an erection. The erection will usually occur in seven to ten minutes. The active ingredient in the pellet is alprostadil, which is also used for injection into the penis. MUSE avoids the use of a needle and syringe and has minimal pain or discomfort with its use. Men will usually get an erection that will last for forty-five to sixty minutes. The actual duration will vary from patient to patient. The pellets can be used up to two times a day. The drug is effective in sixty-five percent of the men who have used this treatment option.

The most common side effect is a dull ache at the tip of the penis, which occurs in only a few of the men that use the pellet. Most men report that with repeated use of the pellet the discomfort lessens. If the man has a partner in the childbearing age group, we suggest that the man use a condom to prevent the drug from being absorbed by the female partner. Rarely, the erection will last several hours and result in priapism which we have described as a medical emergency. Priapism requires the use of other medications that will reverse the effect of the MUSE.

MUSE comes in four strengths—125, 250, 500-hundred, and 1,000 micrograms. The doctor must select the correct dose of the MUSE to insert that will produce the desired duration and quality of an erection and avoid any of the side effects or complications.

Third-Line Treatment

When first-line and second-line treatments are not effective and a man still desires to have sexual intimacy with his partner, then surgery or third-line treatment is an option.

A penile prosthesis is the surgical insertion of device through a very small, two-inch incision, which allows a man to create

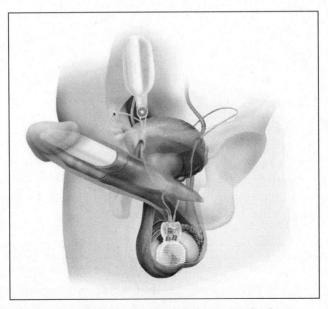

Figure 2: Inflatable penile prosthesis *(Coloplast)*

an erection whenever and for as long as he wants. The most popular is the inflatable penile prosthesis, which was developed in 1973. The inflatable prosthesis consists to two cylinders, which are inserted into the penis, a small pump inserted into the scrotum, and a reservoir of a harmless salt solution place behind the muscles of the abdomen.

When the man desires to have an erection, he squeezes the pump located in the scrotum and fluid is moved from the reservoir behind the muscle of the abdomen into the cylinders in the penis allowing the penis to increase in girth and length. This produces a natural feeling erection.

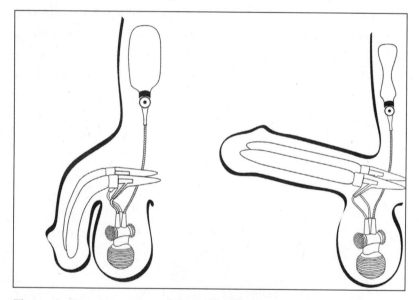

Figure 3: Demonstration of the Inflatable Prosthesis in the Flaccid Position and in the Erect Position *(Coloplast)*

After engaging in sexual intimacy and when he desires for the penis to become soft, the man compresses a release valve located on the pump, and the fluid reverses its direction and returns to the reservoir causing the penis to become flaccid. The entire device is completely concealed and cannot be detected even under the closest scrutiny. Men with the prosthesis can comfortably shower and change clothes in the men's locker room without anyone being able to suspect that a prosthesis is in place.

The complications of the inflatable prosthesis include bleeding, infection and failure. Rarely is bleeding or infection a problem. The device is successful in ninety-eight percent of patients and seldom has to be repaired or replaced. The patient and the partner have nearly one-hundred percent satisfaction with the device. The procedure is done in the hospital or one-day stay surgical center. Most men are discharged within a few hours after the procedure, or after they are able to urinate. Most men can begin using their prosthesis within four weeks after the procedure. Most insurance companies, including Medicare, pay for the procedure.

A malleable implant consists of two semi-rigid rods which are surgically inserted through a small incision into the penis.

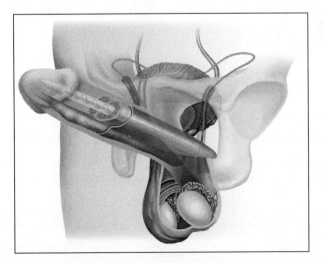

Figure 4: Malleable Penile Prosthesis in the Erect Position *(Coloplast)*

The man can place the penis into the position for intercourse or push it down to conceal the penis in his underwear.

Two other options are low-intensity shock-wave therapy and surgery to correct the problem of abnormal blood supply to the penis. These two treatments are only recommended for men with ED who are resistant to other treatments. There are a few studies that suggests that shock waves are effective in men who have decreased blood supply to the penis and are unresponsive to oral medications, the phosphodiesterase-five inhibitors.

In cases when vascular reconstructive surgery is performed to improve the blood supply of the penis, blocked arteries are bypassed by transferring an artery from an abdominal muscle to a penile artery. Only a very small number of men would benefit from revascularization, and it is not commonly performed because of the level of difficulty, high cost, and suboptimal success rate. It may be suggested to men who have experienced some trauma around the scrotum and anus regions. Additionally, vascular surgery may be an appropriate treatment for men who are born with veins that have difficulty draining blood out of the penis, which is a very uncommon condition.

The Bend with No End—Treatment of Peyronie's Disease (PD)

PD is a health issue in which a scar or plaque forms under the skin of the penis. Most of this plaque builds up leading to a curved erection, which can make intercourse difficult or painful for a man and his partner. PD is thought to impact about six percent of men between the ages of forty and seventy years of age. This may be an underestimate, as many men are too embarrassed to seek help or report the problem to their physician.

Doctors do not know what causes PD or how to prevent the disease before it starts. We do know that PD runs in the family. We also know that forceful sexual activity causes minor injuries to the penis that may initiate the process of scar formation. The incidence is also slightly higher in men who have undergone surgery or radiation for the treatment of prostate cancer. The disease is not related to a sexually transmitted disease or cancer of the penis.

The symptoms of PD include a curvature of the penis. A man may also be able to feel one or more hard lumps on the penis. Men may also experience painful erections and the man's partner may also complain of vaginal pain or discomfort. Peyronie's disease may also lead to erectile dysfunction for several reasons: blockage of blood flow by the plaque, pain from the scar tissue, or the emotional effect from the curvature.

The treatment options for PD include watchful waiting, non-surgical treatments, and surgery. Watchful waiting is appropriate for a man with a minimal curvature that does not cause pain or impede sexual intimacy. Injection therapy includes the use of interferon, verapamil, or collagenase drugs (Xiaflex) directly into the plaque, which may dissolve the plaque.

Surgical options include straightening surgery, referred to as plication. The plaque can also be removed, followed by repair of the resulting defect with a graft of human tissue or synthetic tissue. Finally, the plaque can be surgically removed and a penile prosthesis inserted, thereby removing the offending bend in the penis.

Peyronie's disease is a common urologic disorder. It is a source of potential pain and discomfort. Help is available and most men can achieve relief from the disabling curvature. Remember, this is not the type of curve you want on your partner!

Alternative Treatment Options

There are going to be some men who opt not to use medications or consider injections or surgery for solving their ED problem. Alternative options include hypnosis, acupuncture, meditation or mindfulness, ginseng, and DHEA supplements.

Acupuncture is an ancient Chinese practice in which an acupuncturist inserts six to ten tiny needles into specific sites along the body's channels or "meridians." Treatment generally occurs once per week for ten to twenty weeks.

Meditation or mindfulness is a technique of training the mind to focus on the present. Meditation is a way of reducing stress and achieving a sense of inner peace that can benefit both emotional health and overall physical condition. Types of meditation include deep breathing, visualized breathing, progressive muscle relaxation, and guided imagery.

In the case of erectile dysfunction, meditation is usually most effective when the condition is not caused by a physical issue that restricts the flow of blood to the penis but is psychological in origin.

Ginseng (or Panax ginseng) is sometimes referred to as "herbal Viagra." The root of the ginseng plant has been used in traditional Chinese medicine for thousands of years. Red ginseng contains natural antioxidants, which are known to reduce inflammation, boost energy levels, and help the body deal with fatigue. These antioxidants may also improve blood flow through the body, including blood flow to the penis.

DHEA, or dehydroepiandrosterone, is the most common circulating steroid hormone in the human body. Men with erectile dysfunction often have low levels of DHEA, so it is logical to think that taking DHEA supplements might protect against erectile dysfunction. Some people even believe that DHEA is a "fountain of youth" and can slow down the aging process.

Alternative treatments are generally safe, inexpensive, and certainly will not make the ED worse. However, keep in mind that most alternative treatments are not covered by insurance. Most importantly, erectile dysfunction may be a sign of an underlying medical condition, so a complete history, physical examination, and a few laboratory tests are suggested.

Another alternative treatment option is low intensity shock wave therapy. This was first used in the 1980s for the treatment of kidney stones. Now this same energy source is used to enhance the blood supply to the penis and help restore erections.[4] Treatment consists of twelve weeks of shock-wave sessions, after which a man with ED ideally doesn't need to worry about any treatment for at least two years.[5]

Treatment for Psychogenic ED

For men with psychogenic ED, counseling may alleviate the problem. Often times the man and his partner will attend counseling sessions, especially if there is discord in the relationship between the man and his partner. Psychotherapy with a counselor is often very effective for men who have performance anxiety as a cause of their ED.[6] Often times men with psychogenic ED can have their erections jump-started with oral phosphodiesterase-five inhibitor medications (Viagra, Levitra, or Cialis), which are used

---

4. Gruenwald, I., Appel, B., Vardi, Y. "Low-intensity Extracorporeal Shock Wave Therapy—A Novel Effective Treatment for Erectile Dysfunction in Severe ED Patients who Respond Poorly to PDE% Inhibitor Therapy." *The Journal of Sexual Medicine*, no. 9 (2012): 259–264.

5. Schultz, R. "Is This the New Viagra?" *Men's Health*, September 10, 2014.

6. Kaplan, HS. The New Sex Therapy, Routledge, New York 2012.

temporarily until the man gains confidence and control of his erections. He can then discontinue the use of the medications.

There are some clues that you can use to give you an indication that you may have a psychogenic component to your ED. Men with premature ejaculation and ED may have psychogenic element to their ED (see Chapter 5). If a man has ED that is partner specific, that is ED with one partner and not with other partners, it is a strong suggestion of psychogenic ED. Finally, if a man masturbates and achieves a full erection with normal ejaculation, then this too suggests psychogenic ED.

One of the most common causes of psychogenic ED is performance anxiety. Performance anxiety appears usually in young men after they fail to achieve an erection. The man will lose his self-confidence the next time he attempts to participate in sexual intimacy. Anxiety provokes an increase in adrenalin levels in the blood, which is a hormone that constricts the blood supply to the penis and makes having an erection difficult. The negative emotions of these events are stored in the brain and repeat themselves every time a man with performance anxiety attempts sexual intercourse. Unfortunately, the problem worsens each and every time the man attempts sexual activity, thereby creating a vicious cycle.

Once psychogenic ED has been diagnosed, help is necessary to reduce anxiety. Counseling to overcome performance anxiety by a professional is very important in these situations.

We suggest that if you are going to embark on sex counseling, check the credentials of the therapist and ask for references. Perhaps the therapist will even arrange for you to speak to a man who has successfully completed sex therapy. This isn't as difficult as you might think, as men who have been helped are often willing and sometimes eager to speak to another man who is suffering from psychogenic ED. We think it's important to look for a focus on human sexuality or the words "sex therapist" in the professional's title. Professionals often hold degrees in marriage and family therapy, social work, theology, psychology, or medicine, but you want to be in the hands of a certified sex therapist who has experience treating psychogenic ED.

What Happened to Mr. Johnson's Johnson?

Mr. Johnson went to his primary care physician who arranged for him to see a nutritionist and started him on a diet program. He lost thirty-five pounds in four months. He also started an exercise program with a trainer, and as a result of the weight loss and exercise program, his glucose and hemoglobin A1C became normal. His cholesterol and blood pressure also returned to normal. He was provided with a prescription for Viagra, fifty milligrams, and had a response after using the medication on the second try. He found that he didn't need the medication after the weight loss. He reports that his marriage is back on track, he has more energy, an enhanced libido, and improvement in his sexual performance. Both Mr. and Mrs. Johnson are living happily ever after!

Bottom Line

Erectile dysfunction is a common condition affecting millions of American men. The diagnosis is easily made. Nearly every man with this problem can be helped. Today, no one needs to suffer the "tragedy of the bedroom." Call your doctor for an evaluation and treatment; you will be much happier and so will your partner!

CHAPTER 5

EJD or Ejaculatory Dysfunction—The Other Major Sex Problem "Hold on, I'm Cummin'!"

Jayden is a thirty-five year-old man with a long history of rapid ejaculation. It is a source of great embarrassment and is causing friction between him and his wife. They are also interested in having children, and Jayden's wife believes that the problem with his ejaculation is contributing to their problem of a barren marriage.

Jayden is not alone. Approximately thirty percent of American men over age eighteen suffer from premature ejaculation.[1] Problems with ejaculation are more common than erectile

---

1. Barnes, T. and Eardley, I. "Premature Ejaculation: The Scope of the Problem." *Journal of Sex and Marital Therapy* 33, no. 3 (2007): 151–170.

dysfunction (impotence). It is a problem that is associated with great embarrassment and a contributor to disharmony between partners. This chapter will define premature ejaculation, discuss the causes, and offer treatment suggestions that can resolve this common problem that affects so many men. We will also discuss delayed (retarded) ejaculation and retrograde ejaculation (backwards flow of fluid into the bladder during ejaculation).

Most people think that sexual dysfunction pertains only to erection problems or impotence. However, problems with ejaculation are more common than erectile dysfunction. This chapter will discuss how ejaculation occurs, what are the most common ejaculatory problems, and what treatments are available for this common condition.

How Does Ejaculation Work?

During normal ejaculation, the fluid goes forward through the urethra (the tube in the penis that transports both semen and urine to the outside of the body). At the same time, internal sphincter (pronounced sphink-ter) muscles close off the opening of the bladder to prevent semen from entering the bladder. This function of the internal muscles is necessary so that the seminal fluid moves outside the body, and does not go backward in a retrograde fashion, or mix with urine making the possibility of conception and pregnancy impossible. This will be explored in greater detail below under retrograde ejaculation.

When you're aroused, tubes called the vas deferens contract, and the sperm in the tubes from the testes move the sperm toward the back of the urethra. At the same time, the seminal vesicles also release their fluids to mix with the sperm (Figure 1). The urethra senses the sperm and fluid mixture. Then, at

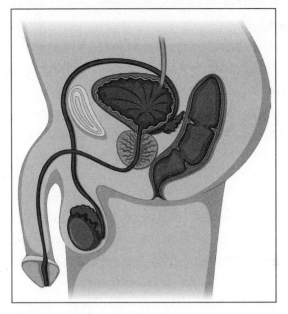

Figure 1: Male reproductive anatomy
(Shutterstock)

the height of sexual excitement, it sends signals to your spinal cord, which in turn sends signals to the muscles at the base of your penis. These muscles contract powerfully and quickly, every eight-tenths of a second. This forces the semen out of the penis as a man reaches his climax.[2]

What is Male Orgasm?

An orgasm consists of three steps: arousal, orgasm, and resolution. Perhaps the largest sex organ in a man is not between his legs but between his ears, i.e., the brain. When a man becomes

---

2. "How the Penis Works: Erection and Ejaculation." WebMD. http://www.webmd.com/erectile-dysfunction/how-an-erection-occurs.

sexually aroused, the brain sends an electrical message down the spinal cord to the nerves that supply the penis, prostate, and testes. With arousal there is opening or dilating of the blood vessels to the penis. During this process, more blood rushes into the penis than leaves the penis, and the penis becomes erect. There is also a contraction of the muscles of the scrotum, which pulls the scrotum towards the body, and the muscles throughout the body increase in tension.

This arousal stage lasts from a few seconds to a few minutes. Prior to ejaculation, a clear fluid is deposited into the urethra or the tube in the penis that transports semen and urine from inside the body to the outside of the body. This pre-ejaculatory fluid is meant to improve the chances of sperm survival.

The orgasm itself occurs in two phases—emission and ejaculation. In emission, the man reaches ejaculatory inevitability, often called the "point of no return." We like to think of ejaculatory inevitability as that point if you are in bed with your partner, and there's a knock at the door stating the President of the United States wants to meet with you, you would be so involved in sexual intimacy that you wouldn't hear the knock. Even if you did hear the knock, you would continue engaging in sexual intimacy and ignore your important visitor! The point of no return can also be compared to a sneeze, which cannot be aborted once the spinal reflex is initiated.

The seminal fluid contains prostate fluid, fluid from the ejaculatory ducts, and sperm. It is deposited near the top of the urethra, ready for ejaculation. Ejaculation occurs in a series of contractions of the penile muscles and around the base of the anus. With the release of fluid there is transmission of a message back to the brain that is interpreted as heightened enjoyment or pleasure.

After ejaculation, the penis begins to lose its erection. About half of the erection is lost immediately, and the rest fades soon after several more seconds. Muscle tension diminishes, and the man may feel relaxed or drowsy. Men usually must undergo a refractory period, or recovery phase, during which they cannot achieve another erection. This usually lasts for about thirty minutes but may be much longer as a man ages.[3]

Ejaculation involves coordinated muscular and neurological events that involve deposition of semen in the urine channel (emission) and ejection of the fluid from the urethral meatus (ejaculation proper). Emission is accomplished by contraction of the vas deferens, seminal vesicles, and ejaculatory ducts. This process is under adrenaline control. Ejaculation results from the rhythmic contractions of the muscles around the urethra, which causes the forcible ejection of the ejaculate. Within the spinal

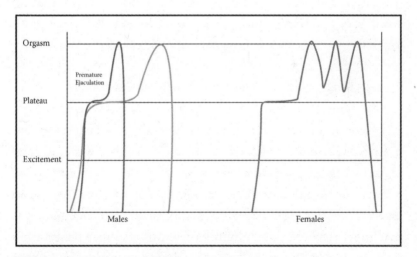

Figure 2: Male Sexual Response Cycle.

3. Thompson, D. Everyday Health. http://www.everydayhealth.com
   /sexual-health/the-male-orgasm.aspx.

cord lies the ejaculation center which is the area involved in the coordination of signals to and from the brain and penis that eventually lead to ejaculation.

The sexual response cycle consists of arousal, plateau, orgasm, and resolution and is shown in Figure 2.

Notice that there is a distinct time difference between male and female sexual response where women require more time to be aroused than do men. Also note that women can be multi-orgasmic, and this rarely occurs in men. The sexual response cycle for men with premature ejaculation is shown in Figure 2. The time to reach orgasm is shorter than normal, with minimal or no time in the plateau phase of the sexual response cycle.

Ejaculatory Dysfunctions

There are four main types of ejaculation: 1) premature ejaculation 2) retrograde ejaculation (3) retarded ejaculation (orgasm) and, (4) failure of ejaculation (anejaculation).

Premature Ejaculation—When It Is Too Quick Down There

Premature ejaculation, also known as rapid ejaculation, lacks a definition that is agreed upon by all physicians but essentially is the condition whereby a patient ejaculates with minimal sexual stimulation, before he wishes it to occur. The time from vaginal penetration to ejaculation in the average American man is nine minutes. The definition of premature ejaculation is reaching orgasm in one minute or less after vaginal penetration (intercourse), or ejaculation that occurs too early for partner satisfaction. This problem is common, occurring in thirty percent of adult men at some time in their lives, and it is the most common form of male sexual dysfunction. It can be caused by erectile dysfunction, anxiety and nerve hypersensitivity and is, in general, quite treatable. There are numerous theories as to the cause, but most cases are probably multi-factorial with a contribution from both psychological and physical factors. The management of this problem is best handled with a combination of psychotherapy and medication.

Brief History of Premature Ejaculation

The phenomenon of premature ejaculations is probably as old as humanity. Writings as early as Greek antiquity made mention of an *ejaculatio ante portas*. But it was not until the late nineteenth century that the experience was described in the medical literature and concerned as a medical disorder or a medical condition. Premature ejaculation has also been called early ejaculation, rapid ejaculation, rapid climax, premature climax, and (historically) ejaculation praecox.

Simply defined, rapid ejaculation or premature ejaculation is uncontrolled ejaculation either before or shortly after sexual penetration, with minimal sexual stimulation, before the man wishes, and with distress to the man and often to his partner. Premature ejaculation is a common sexual complaint. Estimates vary, but as many as one out of three men say they experience this problem at some time during their lives. Premature ejaculation is the most common sexual problem occurring in men under the age of forty. As long as it happens infrequently, it's not cause for concern. However, when it occurs all or the majority of the time, it is a source of anxiety, tension, and can contribute to discord between a man and his partner.

However, a man may meet the diagnostic criteria for premature ejaculation if:

- Always or nearly always he ejaculates within one minute of penetration.
- He is unable to delay ejaculation during intercourse all or nearly all of the time.
- He feels distressed and frustrated and tends to avoid sexual intimacy as a result.

There are two classifications of premature ejaculation: lifelong (primary) or acquired (secondary). Lifelong premature ejaculation occurs all or nearly all of the time beginning with your first sexual experience. Acquired premature ejaculation has the same symptoms but develops after you've had previous sexual experiences without ejaculatory problems.

Not all men can fit into these two categories. Many men feel that they have symptoms of premature ejaculation but the symptoms do not meet the diagnostic criteria for premature ejaculation. Instead they may have variable premature ejaculation, which is characterized by periods of rapid ejaculation as well as periods of normal ejaculation. For example, a man may have premature ejaculation with one partner and not with another partner, nonetheless still causing him great consternation.

Causes of Premature Ejaculation

The exact cause of premature ejaculation isn't known. While it was once thought to be only psychological in origin, doctors now know that premature ejaculation is more complicated and involves a complex interaction of psychological and biological factors.

Psychological Causes

Early sexual experiences may establish a pattern that can be difficult to change later in life. For example, there are situations in which you may have hurried to reach climax in order to avoid being discovered in the act of sexual intimacy. This is common among young men with their first few sexual encounters. The rapid ejaculation becomes a pattern and the brain is trained to complete the act quickly. Later in life, this pattern is difficult to alter and the premature ejaculation continues.

Another example is a man who has received advice that sexual intercourse should not take place until after marriage. If the man participates in sexual intimacy, there may be guilt associated with intercourse and may increase a man's tendency to rush through sexual encounters.

Other factors that can play a role in causing premature ejaculation include:

- Erectile dysfunction: Men who are anxious about obtaining or maintaining an erection during sexual intercourse may form a pattern of rushing to ejaculate, which can be difficult to change. Also, when a man struggles to stay stimulated enough to maintain his erection, he may hasten the time to reach climax.
- **Anxiety:** Many men with premature ejaculation also have problems with anxiety—either specifically about sexual performance or related to other issues.
- **Relationship problems:** If you have had satisfying sexual relationships with other partners in which premature ejaculation happened infrequently or not at all, it's possible that interpersonal issues between

you and your current partner are contributing to
the problem.

Biological Causes

Psychological factors are the most common cause of premature ejaculation. However, there are a number of biological
factors that may contribute to premature ejaculation.

The most common biological cause is an infection of the
prostate gland and/or the urethra. Inflammation of the prostate gland can make the prostate more sensitive and trigger a
rapid ejaculation. Abnormal levels of brain chemicals called
neurotransmitters, such as serotonin, may be associated with
premature ejaculation. Also, too much thyroid hormone or
hyperthyroidism may contribute to premature ejaculation.

Other causes of PE related to certain medications and also
associated with the use of alcohol and recreational drugs need
to be evaluated.[4]

There is also a possibility that premature ejaculation can be
an inherited problem, but it is difficult for most men to obtain
information from their father or grandfather about their sexual
experiences. Finally, nerve damage from surgery or trauma is a
rare cause of premature ejaculation.

We have listed the most common causes of premature ejaculation. However, there are other factors that can increase your
risk of premature ejaculation.

You may be at increased risk of premature ejaculation if
you occasionally or consistently have trouble getting or maintaining an erection. Fear of losing your erection may cause you

---

4. Pugh, J. and Belenkos, S. "Alcohol, Drugs and Sexual Function: A
   Review." *Journal of Psychoactive Drugs*, 2001: 223–233.

to consciously or unconsciously hurry through sexual encounters (for more information on erectile dysfunction, see Chapter 4).

If you have a serious or chronic medical condition, such as heart disease, you may feel anxious during sex and may unknowingly rush to ejaculate. Often a discussion with your doctor about engaging in sexual intimacy can alleviate those fears and concerns about engaging in sexual intimacy.

Emotional or mental strain in any area of your life can play a role in premature ejaculation, often limiting your ability to relax and focus during sexual encounters. Engaging in sexual activity requires a certain level of relaxation to make the nervous system function properly. Stress can be a distraction and fails to allow proper function of the nerves that supply the penis and the prostate gland, and stress can create a situation that results in premature ejaculation.

Other Manifestations of Premature Ejaculation

While premature ejaculation alone doesn't increase your risk of health problems, it can cause significant problems in your personal life and can be a source of personal distress not only to the man but also to his partner. Men who are not in a relationship appear to have the greatest distress, as the condition often precludes the man from seeking out or initially maintaining a relationship.

Partners often complain that the man with premature ejaculation is focusing on his performance and not on his partner. Significant numbers of partners report that the man's premature ejaculation resulted in a failure of the relationship.[5]

---

5. Burri, A., Giuliano, F., McMahan, C., Porst, H. "Female Partner's Perception of Premature Ejaculation and its Impact on Relationship Breakups, Relationship Quality, and Sexual Satisfaction." *The Journal of Sexual Medicine*, 2014: 2243–2255.

Probably the most important and common side effects of premature ejaculation are embarrassment, anxiety, stress in the relationship, and even depression.[6]

Premature ejaculation can occasionally make fertilization difficult or impossible for couples that are trying to achieve a pregnancy. Bobby's history is not unusual, and premature ejaculation can result in failure of the sperm to be properly deposited in the vagina and thus prevent the sperm from reaching the woman's egg and allowing natural fertilization to take place.

The Evaluation of Premature Ejaculation

In addition to asking about your sex life, your doctor will ask about your health history and may perform a general physical exam. The health history will consist of questions similar

---

Table 1

Questions you might be asked by your doctor evaluating premature ejaculation.
1. What is the estimated time from penetration to ejaculation?
2. When was your first sexual experience?
3. Do you have an adequate erection which enables to make penetration?
4. Is the problem specific to one partner specific?
5. Do you have problem with rapid ejaculation during masturbation?
6. What is the impact of your problem on your partner?
7. What previous treatment(s) have you had?
8. Is the problem causing anxiety, embarrassment, depression?
9. Is the problem associated with any other medical conditions?

---

6. Matthew, R.J. and Weinman, M.L. "Sexual Dysfunctions in Depression." *Archives of Sexual Behavior*, 1982: 323–328.

to those shown in Table 1. Your doctor may order a urine test to rule out possible infection and a prostate exam to see if the prostate gland is the culprit. If you have both premature ejaculation and trouble getting or maintaining an erection, your doctor may order blood tests to check your male hormone (testosterone) levels or other blood tests to test, for example, your thyroid level.

Treatment of Premature Ejaculation

Common treatment options for premature ejaculation include behavioral techniques, topical anesthetics, oral medications, and counseling. One or more treatment options are often required to treat premature ejaculation.

DIY Solutions For Premature Ejaculation

In some cases, therapy for premature ejaculation may involve taking simple steps, such as masturbating an hour or two before intercourse so that you're able to delay ejaculation during sex. Your doctor also may recommend avoiding intercourse for a period of time and focusing on other types of sexual play so that pressure to perform sexually is removed from your sexual encounters.

The Squeeze That Pleases

Your doctor may instruct you and your partner in the use of a method called the pause-squeeze technique. This technique, which was developed by Masters and Johnson, pioneering sex therapists in the 1960s, pushes blood out the penis and momentarily decreases sexual tension and represses the ejaculatory response. It has been successful in some men with premature ejaculation.[7]

The pause-squeeze technique is easy to learn. You begin sexual activity as usual, including stimulation of the penis by yourself or by your partner, until you feel almost ready to ejaculate. When you feel close to having an ejaculation, have your partner squeeze the end of your penis, at the point where the head (glans) joins the shaft, and maintain the squeeze for several seconds, until the urge to ejaculate passes. Usually a five to seven second squeeze is adequate. After the squeeze is released, wait for about thirty seconds, and then resume foreplay. You may notice that squeezing the penis causes it to become less erect, but when sexual stimulation is resumed, it soon regains a full erection.

By repeating this as many times as necessary, you can reach the point of entering your partner without ejaculating. After a few practice sessions, the feeling of knowing how to delay ejaculation may become a habit that no longer requires the pause-squeeze technique.

Masturbate Before You Copulate

This is a time-tested treatment that is effective as well as inexpensive. One caveat, as a man ages, the time from one ejaculation to the next erection, referred to as the refractory period, increases in duration. That means you'll need to allow yourself more time between masturbation and sexual intimacy than you did back in your teen years. Failure to follow this advice may result in trading your premature ejaculation for impotence,

---

7. Masters, W.H. and Johnson, V. *Human Sexual Response*. Little Brown, 1966.

which is no less embarrassing or discouraging. How do you find out? You have to experiment and find that precious time interval that works from time of masturbation to intercourse.

Try not to sweat the small stuff. Just putting pressure on yourself may be the culprit causing using your PE. The likelihood of having simultaneous orgasms is rare as hen's teeth. It just doesn't happen the majority of the time. You can allay lots of apprehension if you just relax and enjoy the moment of sexual intimacy without the pressure of having an orgasm at the same time as your partner.

Distract yourself and don't think about your orgasm. If you notice yourself getting too excited, turn your thoughts to something distant, abstract, and unsexy, such as counting sheep, subtracting seven from one hundred down to zero, or the anticipated scores of your favorite football team. Avoid thinking of a topic that is going to make you stressed, like your quota in business or your bonus, as that may cause you to lose your erection.

Getting the Upper Hand

Orgasm control is the practice of maintaining a high level of sexual arousal while delaying ejaculation. It takes practice, but it gets easier over time. Here are methods recommended by the National Institutes of Health to stop premature ejaculation:

Stop-and-start method. Have intercourse as usual until you feel yourself coming uncomfortably close to orgasm. Immediately and abruptly cease all stimulation for thirty seconds, then start again. Repeat this pattern until you're ready to ejaculate.

Change it up. Some intercourse positions put less pressure on the glans. Try "passive" positions. Go from missionary or male superior to male on the bottom. Missionary and rear-entry positions place the most stimulation and friction on the glans, so consider taking them off the table until more control is achieved.

Go slow. Depending on your personal sensitivity, slowing your movements and pelvic thrusting and opting for gentler intercourse can help you hold off orgasm. If you find yourself getting too close to orgasm, put the brakes on, change to a new position, or take a pause to stimulate your partner in other ways. It just may be the pause that refreshes.

Focus on foreplay. Never forget that you can give your partner a great sexual experience through more extended, intimate, attentive, and generous foreplay. Stimulate your partner enough manually, orally, or with toys such as vibrators, and they may not need a prolonged period to achieve sexual satisfaction.

A Kegel a Day May Keep Ejaculation at Bay

Help may be just a Kegel away. Kegel exercises have been recommended to women for decades to control incontinence (the urge to urinate all the time) and to enhance sexual responses during intercourse. Now the Kegel exercise has been demonstrated to be effective for men. Kegel exercises are used to strengthen the muscles responsible for bladder control. These pelvic floor muscles are the same muscles you use to prevent urination, or to keep from passing gas from your rectum at an inappropriate moment.

There's no better method to strengthen the pelvic region in both men and women than to create a strong pubococcygeus muscle (PC muscle), which can help control ejaculation. The easiest way for a man to find this muscle is to see if he can stop the flow of urine when going to the bathroom. It's the PC muscle that you use to do that. Once you find it, you need to practice feeling exactly where it is located and make sure you can contract and relax your PC muscle, rather than using your abdominals, buttocks, or thighs (these must all stay loose when doing Kegel exercises). To do Kegels, you will clench and release the PC muscle repeatedly for a slow count to five each time and then relax for ten seconds. If you do the Kegel exercise ten times, that is one set. Perform one set of ten reps, three times per day (thirty total per day). The Kegel exercise can then be performed during sexual intimacy allowing your PC muscle to prevent orgasm. After two to three weeks of exercising, you should see improvement. Remember, Arnold Schwarzenegger did not become an Olympic champion after his first visit to the weight lifting room. The same applies to the man with premature ejaculation. One day of exercises is not going to result in significant improvement in building up the PC muscle. The reality is that it may take several weeks, or even months, to see improvement in delaying ejaculation.

Condoms are not just for prevention of sexual transmitted diseases (STDs). Something as simple as an over-the-counter or machine-generated condom works for a lot of men with PE. Condoms decrease the stimulation of the very sensitive glans, or head of the penis, for most men, which will often prolong the time before ejaculation.

Some condoms are coated with a slight numbing gel on the inside. This can help you prolong ejaculation without causing numbness to your partner. If you opt for this technique, just make sure you know which side is where when you put it on because the wrong side puts your partner on the wrong side of the orgasm track.

Topical Anesthetics: Putting the Numb On Mr. Happy

Anesthetic creams and sprays that contain a numbing agent, such as lidocaine or prilocaine, are often used to treat premature ejaculation. These products, which can be obtained over the counter and through the Internet, are applied to the penis a short time before sexual intimacy to reduce sensation and thus help delay ejaculation. Unfortunately, some of these preparations may be transferred to the partner, thus making sexual enjoyment for the partner diminished or absent because of numbness of the vagina and particularly the clitoris. The only solution is to apply the topical anesthetic and then put on a condom to avoid transfer to the partner.

A lidocaine spray for premature ejaculation (Promescent) is now available over-the-counter. Promescent is a topical medication that is applied to the penis ten minutes before sexual activity and helps a man to better manage the sensations of sex through desensitization. However, unlike other topical medications for early ejaculation, Promescent penetrates only the most superficial layer of skin on the head of the penis, or the glans of the penis, and is not absorbed through the skin on the shaft of the penis. Since it is not absorbed into the skin, it will not negatively impact his partner's sensations or result in decreased sensation for the partner.

Although topical anesthetic agents are effective and well-tolerated, they have potential side effects. For example, some men report temporary loss of sensitivity and decreased sexual pleasure. In some cases, female partners also have reported these effects of lack of sensation. In rare cases, lidocaine or prilocaine can cause an allergic reaction.

Oral Medications

Many oral medications may delay orgasm. Although none of these drugs is specifically approved by the Food and Drug Administration to treat premature ejaculation, some are used for this purpose, including antidepressants, analgesics, and phosphodiesterase-five inhibitors, which include Viagra, Cialis, and Levitra. These medications may be prescribed for either use thirty to forty-five minutes before sexual intimacy or daily use, and may be prescribed alone or in combination with other treatments.

Many years ago it was noted that one of the side effects of certain antidepressants is delayed orgasm. For this reason, selective serotonin reuptake inhibitors (SSRIs), such as sertraline (Zoloft), paroxetine (Paxil) or fluoxetine (Prozac, Sarafem), are used to help delay ejaculation. If SSRIs don't improve the timing of your ejaculation, your doctor may prescribe the tricyclic antidepressant clomipramine (Anafranil). The dosage of these medications used for treating premature ejaculation is shown in Table 2.

Table 2
SSRI medications used to treat premature ejaculation.

Drug (Trade name)	Dosage before intercourse	Side effects
Clomipramine (Anafranil)	25-50mg	nausea
Paroxetine (Paxil)	20mg	sleepiness, yawing
Sertraline (Zoloft)	50mg	headache, fatigue
Fluoxetine (Prozac)	5-20mg	headache, insomnia

These drugs can be taken several hours before sexual activity, and since they inhibit arousal, they can help make it easier for a man to control ejaculation. Before he makes a decision regarding these drugs, which require a prescription, he will need to see either his general practitioner or a urologist; he should also ask about the side effects of SSRIs. Unwanted side effects of antidepressants may include nausea, dry mouth, drowsiness, and decreased libido. Also, these medications should not be abruptly discontinued.

Priligy (Pri-Low-Gee) May End the Premature Ejaculation Odyssey

Dapoxetine (Priligy) is the only drug that has been created specifically for the treatment of premature ejaculation by increasing the neurotransmitter serotonin, which can be effective in prolonging the time from penetration to ejaculation. The drug has been approved in over fifty countries for the treatment of premature ejaculation but has not received approval from the FDA at the time of publication of this book. The thirty to sixty mg tablets, taken one to two hours before sexual intimacy, have shown to be effective, even at the first dose, with an improvement of the time from penetration to ejaculation. The drug has been reported to be effective in men with life-long premature ejaculation as well as men who acquired premature ejaculation later in life. Also, the drug has been effective in men with premature ejaculation and erectile dysfunction treated with any of the phosphodiesterase-five-inhibitors (Viagra, Levitra, Cialis). The side effects occur in less than three percent of the men using dapoxetine and consist of transient nausea, dizziness, and diarrhea.

Tramadol (Ultram) is a medication commonly used to treat pain. It also has side effects that delay ejaculation. Tramadol may be prescribed when SSRIs haven't been effective. Unwanted side effects of tramadol may include nausea, headache, and dizziness. How tramadol works has not been completely elucidated. Perhaps the mechanism of action is the anesthetic-like effect. Because of the potential of tramadol to cause addiction, its off-label usage has been discouraged. However, tramadol may be considered when other treatments have failed.

Hold 'Em Up and Keep 'Em Up—Using ED Drugs to Treat Premature Ejaculation

Some medications used to treat erectile dysfunction, such as sildenafil (Viagra, Revatio), tadalafil (Cialis, Adcirca) or vardenafil (Levitra, Staxyn), also may help premature ejaculation. By enhancing erectile function, less focus can be placed on maintaining an erection, thereby reducing the possibility of overstimulation. Unwanted side effects may include headache, facial flushing, temporary visual changes, and nasal congestion.[8]

These oral medications can be used alone for treating ED or in combination with the SSRIs as a treatment for premature ejaculation. These ED drugs are particularly effective in men who have both ED and premature ejaculation. It is worth noting that nearly fifty percent of men with ED also have a problem with premature ejaculation.[9]

Alpha-blockers to Unblock Premature Ejaculation

Older or middle aged men with urinary symptoms such as frequency of urination, getting up at night to urinate, and dribbling after urination as a result of the enlarged prostate gland who also have premature ejaculation, may be effectively treated with drugs that block the nerves to the prostate gland muscles that control ejaculation. This is a nice example of a "two-fer" or one drug used to treat two conditions. However, these drugs can also cause retrograde ejaculation as a side effect (discussed later in this chapter).

---

8. McMahon, C.G., Stuckey, B., Andersen, M.L. et al. "Efficacy of Sildenafil Citrate (Viagra) in Men with Premature Ejactulation." *The Journal of Sexual Medicine*, 2005: 368–375.
9. Jannini, E.A., Lombardo, F., Lenzi, A. "Correlation Between Ejaculatory and Erectile Dysfunction." *International Journal of Andrology*, 2005: 40–45.

Drugs such as tamsulosin (Flomax), siladosin (Rapaflo) or alfuzosin (Uroxatrol) are alpha-blockers and have been used for decades for treating the enlarged prostate gland (see Chapter 2 on prostate gland enlargement). The mechanism of action is that these drugs effectively block the nerve transmission of nerve impulses to the prostate gland and not only decrease the resistance to the flow of urine from the bladder to the outside of the body but also decrease the sensitivity to the prostate gland. Ejaculation is a reflex that may be impacted by the use of these alpha-blocker medications. There is documentation that alpha-blockers may be effective in delaying ejaculation in approximately fifty percent of the men using these alpha-blocking drugs.[10]

Psychological Intervention: Mind Over Splatter

This approach, also known as talk therapy, involves talking with a mental health provider, such as a social worker, psychologist, or sex therapist, about your relationships and experiences. These sessions can help you reduce performance anxiety and find better ways of coping with stress. Counseling is most likely to help when it's used in combination with drug therapy. Couples can attend the sessions together, or the counselor may suggest that the man and partner are seen separately.

Counseling with a therapist helps men develop the skills that help them to delay ejaculations while increasing the man's self-confidence and decreasing performance anxiety. Counseling can also help to resolve interpersonal issues that may have precipitated the premature ejaculation or have occurred as a result of the sexual problem. Counseling may occur one-on-one

10. Choi, J.H., Jung Hwa, J.S., Kam, S.C. et al. "Effects of Tamsulosin on Premature Ejaculation in Men with Benign Prostatic Hyperplasia." *The World J Mens Health*, 2014: 99–104.

with the therapist, with both the man and his partner being treated simultaneously, or in a group environment, with several couples being treated at the same time by the therapist. It has been our observation that a man who is not in a relationship is more difficult to treat than a man who is in a relationship and has a sexual partner. It is also our experience that premature ejaculation is more effectively treated in conjunction with medical therapy than with counseling alone.

Alternative Treatments for Premature Ejaculation

Several alternative medicine treatments have been studied, including selected behavioral therapies (such as yoga). Additionally, surgical treatments (such as surgically dividing the nerves to the head of the penis to decrease its sensitivity) have been done, but only on men who have been refractory to conservative treatment with medication and/or counseling. The role of surgery in the management of premature ejaculation remains unproven and until the results of further studies have been reported, we do not recommend this approach to treatment.

Yoga is a popular form of complementary and alternative treatment. Use of yoga for various bodily ailments is recommended in ancient *Ayurveda* (ayus = life, veda = knowledge) texts and is being increasingly investigated scientifically. Many patients and yoga protagonists claim that it is useful in men with premature ejaculation. Several studies have documented that yoga appears to be a feasible, safe, effective, and acceptable non-pharmacological option for premature ejaculation.

Going Au Naturel—Herbal Remedies and Supplements

Some herbal remedies may help to alleviate the problem. Take one hundred milligrams of kava before engaging in sexual intercourse. Kava root comes in pill form and powder form, which can be brewed into a tea and is sold at many health food stores. Kava increases blood flow to the penis and slows reaction to increased sexual stimulation, therefore helping you maintain erection longer.

Add two drops of hibiscus flower essence to one-quarter cup of water and sip slowly. Hibiscus flower essence helps relieve stress and promotes romantic sexual feelings, which increases the ability to maintain erection and control ejaculation.

Take a multivitamin formulated for men daily. Using multivitamin supplements in conjunction with a healthy diet and exercise may increase sexual stamina and performance.

Take Graphites Twelve C two to four times daily. This nutrient in supplement form will help stop premature ejaculation with repeated use. Once your condition has improved, you may discontinue use of this supplement.

Try taking a 5HTP (hydroxytryptophan) supplement daily. 5HTP is a naturally occurring nutrient that is responsible for sexual health and serotonin production. It suppresses the urge to ejaculate sooner than normal while increasing stamina. You can find it at many health food stores and pharmacies.

Retrograde Ejaculation

Retrograde ejaculation is the process whereby the semen is passed in a retrograde or backward direction into the bladder rather than exiting the urethra. There are three potential causes to this problem: anatomic (following surgery on the prostate gland or from a congenital process); neurologic (due to disorders that interfere with the ability of the bladder neck to close during emission such as diabetes mellitus, or surgery that can affect the nerves to the bladder and prostate gland); and side effects of certain medications that cause paralysis of the internal muscles or sphincters. This condition is diagnosed by the finding of seminal fluid and/or sperm in a urine specimen obtained immediately after orgasm.

The treatment of retrograde ejaculation depends to some extent on the cause. Anatomic causes are rarely curable, and sperm harvesting from the bladder is required for those men with retrograde ejaculation wishing to initiate a pregnancy. Pharmacologic causes are generally reversible by withdrawal of the offending medication. Neurologic causes are difficult to treat if there is complete nerve damage, such as that which may occur in spinal cord injured patients. In those patients with a partial neural injury (diabetes), the use of certain medications (ephedrine and pseudoepohedrine) may convert the man with retrograde ejaculator to an antegrade or normal ejaculator.

Delayed Ejaculation (DE)

A common problem in older men is delayed or retarded ejaculation where it takes more than twenty-five minutes to achieve an orgasm. This can be a source of tension and pressure between a man and his partner. Your partner might feel that you're not attracted to them and that love has gone out the bedroom door. You might feel frustrated or embarrassed about wanting to achieve ejaculation but being physically or mentally unable to do so. Treatment or counseling can help resolve these issues.

The causes of delayed ejaculation include conditions and reactions to medications. Psychological causes of DE can occur due to a traumatic experience. Cultural or religious taboos can give sex a negative connotation. Anxiety and depression can both suppress sexual desire, which may result in DE as well.

Relationship stress, poor communication, and anger can make DE worse. Disappointment in sexual realities with a partner compared to sexual fantasies can also result in DE. Often, men with this problem can ejaculate during masturbation but not during stimulation with a partner.

Certain chemicals can affect the nerves involved in ejaculation. This can affect ejaculation with and without a partner. These medications can all cause DE: antidepressants, such as fluoxetine (Prozac), antipsychotics, such as thioridazine (Mellaril), medications for high blood pressure, such as propranolol (Inderal), diuretics, and alcohol.

Surgeries or trauma may also cause DE. These causes of DE may include the following: damage to the nerves in your spine or pelvis, certain prostate surgeries like prostate gland removal for cancer that cause nerve damage, heart disease that affects blood pressure to the pelvic region, neuropathy or stroke, low thyroid hormone, and low testosterone levels. A temporary ejaculation problem can cause anxiety and depression. This can lead to recurrence, even when the underlying physical cause has been resolved. Also in rare cases, delayed ejaculation is a sign of worsening health problems such as heart disease or diabetes.

The same tests for premature ejaculation are also indicated for delayed ejaculation. Another simple test is to check the reaction of your penis to vibration, which may reveal a physical problem. An easy method of performing this test is to use a tuning fork and check your appreciation of the vibration on your hand compared to the shaft of the penis. If the there is significant decrease in perception of the vibration on the penis compared to the hand, then there is likely a physical explanation of the delayed ejaculation.

At the present time, no drug has been specifically approved for delayed ejaculation, but medications used for conditions such as Parkinson's disease have been shown to help.

Off-label drug use means that a drug that's been approved by the FDA for one purpose is used for a different purpose that has not been approved. Your doctor can still use the drug for that purpose. This is because the FDA regulates the testing and approval of drugs, but not how doctors use drugs to treat their patients. So, your doctor can prescribe a drug however they think is best for you.

Some medications have been used to help DE, but none have been specifically approved for it. Several of these medications include cyproheptadine (Periactin), which is an allergy medication, amantadine (Symmetrel), which is a drug used to treat Parkinson's disease, and buspirone (Buspar), which is an antianxiety medication. Of course, treating illicit drug use and alcoholism, if applicable, can also help DE.

Psychological counseling can help treat depression, anxiety, and fears that trigger or perpetuate delayed ejaculation. Sex therapy may also be useful in addressing the underlying cause of sexual dysfunction. This type of therapy may be completed alone or with your partner.

DE can generally be resolved by treating the mental or physical causes. Identifying and seeking treatment for DE sometimes exposes an underlying medical condition. Once this is treated, DE often resolves. The same is true when the underlying cause is a medication. However, don't stop taking any medication without your doctor's recommendation.

Take home message on delayed ejaculation: Delayed ejaculation does not pose any serious risks to your life; it can, however, be a source of stress and may create problems in your sex life and personal relationships. For most men, help is available.

Whatever Happened to Jayden?

Fortunately, Jayden's wife is very patient and understanding. Their success began simply with conversation. They decided to seek help from a therapist who shared several "practicing" techniques, along with judicious use of a "numbing" agent. With several months of practice, Jayden developed the new "habit" of a more prolonged and satisfying duration of intercourse prior to reaching climax. Their biggest problem will be finding the time for his sexual stamina once the first baby arrives. By the way, they are expecting a new baby!

Bottom Line

Problems with ejaculation are one of the most common ailments to afflict men of all ages. They can wreak havoc on a man's life and certainly impact his partner as well. Help is available, and most men can be helped with both medical and nonmedical management. If you are suffering from ejaculation problems, see your doctor. Premature ejaculation doesn't mean you always have to say, "I'm sorry."

CHAPTER 6

Vasectomy—The Prime Cut

Paco is forty-three years old, has three children, his wife has been on birth control pills for past eight years, and they believe that their family is complete.

Vasectomy is a minor surgery to block sperm from reaching the semen that is ejaculated from the penis. Semen will still exist, but it will contain no sperm. In fact, most of the volume of semen originates from the prostate and two adjacent glands called the seminal vesicles. The testicles contribute only three percent of the total volume of the ejaculate. All of the sperm are contained in this important three percent. After a vasectomy the testes still make sperm, but the sperm are soaked up by the body. Each year, more than half a million men in the U.S. choose vasectomy for birth control. A vasectomy prevents pregnancy better than any other method of birth control, except abstinence. Only one to two women out of a thousand will get pregnant in the year after their partners have had vasectomies.

What Happens Under Normal Conditions?

Both sperm and male sex hormones are made in the paired testes (testicles). The testes are in the scrotum at the base of the penis. Sperm leave the testes through a coiled tube (the "epididymis"), where they stay until they're ready for use. Each epididymis is linked to the prostate by a long tube called the vas deferens (or "vas"). The vas runs from the lower part of the scrotum into the inguinal canal located in the groin. It then goes into the pelvis and behind the bladder. This is where the vas deferens joins with the seminal vesicle and forms the ejaculatory duct. When you ejaculate, seminal fluid from the seminal vesicles and prostate mix with sperm to form semen. The semen flows through the urethra and exits from the end of your penis.

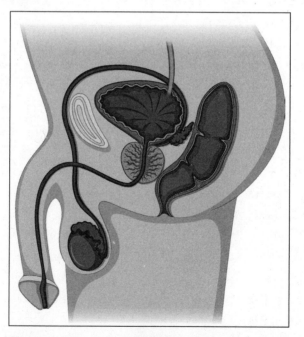

Figure 1: Normal Male Reproductive Anatomy
(Shutterstock)

Vasectomy is the process of dividing the vas (the tube that delivers the sperm from the testis to the prostate) in order to prevent conception. It is the most common method of male contraception in this country. Since vasectomy simply interrupts the delivery of the sperm, it does not change the hormonal function of the testis. Thus, the sexual drive (libido) and ability to engage in sexual intimacy remain intact and will not be affected by the vasectomy. Since most of the semen is composed of fluid from the prostate and seminal vesicles, the semen volume will look the same. Vasectomy is thought to be free of known long-term side effects, and it is considered to be the safest and most reliable method of permanent male sterilization.

The technique of the No-scalpel Vasectomy was developed in 1974 by a Chinese physician, Dr. Li Shunqiang, and has been performed on over eight million men in China.

After injecting the scrotal skin and each vas with a local anesthetic, we use a special vas-fixation clamp to encircle and firmly secure the vas without penetrating the skin. One blade of a forceps or clamp is used to penetrate the scrotal skin. The tips of the forceps are spread, opening the skin much like spreading apart the weaves of fabric. The opening is approximately one-quarter inch in length. The vas is thus exposed and then lifted out and occluded by any of the standard techniques, such as cautery or sutures. The second vas is then brought through the same opening and occluded in a similar fashion. The skin wound contracts to a few millimeters and usually does not require suturing. Patients should feel no discomfort as no needle is used to insert the local anesthetic and the opening is about the size of the tip of a pencil.

Compared to the traditional incision technique, the no-scalpel, no needle vasectomy usually takes less time, causes less discomfort, and may have lower rates of bleeding and infection. There is minimal pain or discomfort as no needle is used to insert the local anesthetic, and no scalpel is used to make the opening in the scrotum. Recovery following the procedure is usually complete in two to three days. Hard work or straining (athletic pursuits or heavy lifting) is not recommended for seven days. Most men should refrain from sexual intimacy for a week after the procedure.

Common Reasons Given for Having a Vasectomy

1. You want to enjoy sex without worrying about pregnancy.
2. You do not want to have more children than you can care for.
3. Your partner has health problems that might make pregnancy difficult.
4. You do not want to risk passing on a hereditary disease or disability.
5. You and your partner don't want to or can't use other kinds of birth control.
6. You want to save your partner from the surgery involved in having her tubes tied and you want to save the expense.

Common Questions Asked About No-Scalpel Vasectomy

How Can I Be Sure That I Want a Vasectomy?

You must be absolutely sure that you don't want to father a child in the future. You should talk to your partner and it certainly is a good idea to make this decision together, consider other kinds of birth control, and talk to friends or relatives who may have had a vasectomy. You and your partner might want to think about how you would feel if your partner had an unplanned pregnancy. Talk to your doctor, nurse, or family planning counselor.

A vasectomy might not be right for you if you are very young, if your current relationship is not permanent, if you are having a vasectomy just to please your partner and you do not really want it, if you are under a lot of stress, or if you are counting on being able to reverse the procedure at a later time.

How Does the Vasectomy Prevent Pregnancy?

Sperm is made in the man's testicles. The sperm then travels from the testicle through a tube called the vas into the body where it enters the prostate gland. In the prostate, the semen is

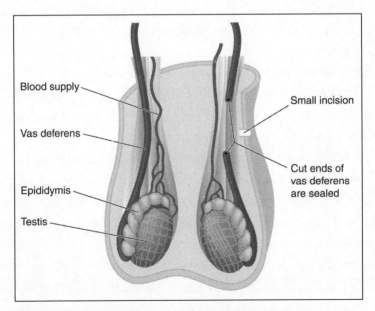

Figure 2: The Vas Before and After the Vas Has Been Divided *(Shutterstock)*

made and here the sperm mixes with the semen. The prostate is connected to the channel in the penis where the sperm and semen are ejaculated into the urethra and then out the end of the penis. In a vasectomy, the vas, or tube, is blocked so that sperm cannot reach the prostate to mix with the semen. Without sperm in the semen a man cannot impregnate his partner. Figure 2 shows the vas before and after the vasectomy.

What Preparation do I Need to Take Before the Vasectomy?

We suggest that you wash the scrotum with any soap to prevent infection. You should bring a pair of tight fitting under-wear\jockey shorts or athletic supporter to the procedure to support the scrotum and minimize swelling after the procedure. We suggest one week prior to the procedure that you avoid using any anti-inflammatory drugs such as aspirin, ibuprofen, and fish oil, as these drugs and supplements may thin the blood and can cause excessive bleeding during or after the procedure.

What is Different About a No-scalpel Vasectomy?

No-scalpel vasectomy is different from a conventional vasectomy in the way that we get to the tubes, or vas, to block them from passing sperm out of the testicles. An improved method of anesthesia helps make the procedure less painful. In a conventional vasectomy, the physician may make one or two small incisions approximately one to one-and-half inches long in the skin with a knife, and the doctor would then use sutures or stitches to close these cuts at the end of the procedure. In the no-scalpel vasectomy, instead of making two incisions, the doctor makes only one very small puncture into the skin with a special instrument. This same instrument is used to gently stretch the skin opening so that the tubes can be reached easily.

The tubes are then blocked, using the same methods as conventional vasectomy. Because a scalpel is not used, there is very little bleeding, and usually no stitches or sutures are needed to close the tiny opening. This opening will heal quickly with little or no scarring. No-scalpel vasectomy was introduced in the United States in 1988 and is now used by many doctors in this country who have mastered the technique.

Reasons for Having a No-scalpel Vasectomy as Compared to Conventional Vasectomy

1. No incision with a scalpel—only a small opening
2. Usually no stitches or sutures will be necessary as the opening is less than 1\4 inch
3. Usually a faster procedure
4. Usually a faster recovery
5. Usually less chance of bleeding and other complications
6. Usually less discomfort
7. Just as effective as regular vasectomy

Will it Hurt?

When the local anesthetic is injected into the skin of the scrotum, you will feel some mild discomfort, but as soon as the anesthetic takes effect you should feel no pain for the rest of the procedure. We recommend that men plan to go home after the procedure and place an ice pack on the scrotum for a few hours to reduce any swelling. Many men will elect to use a bag of frozen peas that will conform to the scrotum, but any ice pack or plastic Zip-Lock bag with a few ice cubes will work just fine. We suggest that the ice pack be placed on top of the underwear and not directly on the skin. We also recommend wearing jockey underwear and not boxer underwear for a few days after the procedure to keep "your package" lightly compressed. Afterward, you will experience mild discomfort for a couple of days and may want to take an over the counter pain killer such as Tylenol, but the discomfort is usually less with the no-scalpel technique because of less trauma or injury to the scrotum and tissues that surround the vas. Also, there are no stitches or sutures in most cases. Your doctor will provide you with complete instructions about what to do after surgery.

How Soon Can I go Back to Work?

You should be able to do routine physical work within forty-eight hours after your vasectomy, and will be able to do heavy physical labor and exercise within a week.

Will the Vasectomy Change Me Sexually?

The only thing that will change is that you will not be able to make your partner pregnant. Your body will continue to produce the same hormones that give you your sex drive and maleness. You will make the same amount of semen. Vasectomy will not change your beard, muscles, sex drive, erections, climaxes, or your even raise your voice! Some men say that without the worry of accidental pregnancy and the bother of other birth control methods, sex is more relaxed and enjoyable than before.

Will I Be Sterile Right Away?

No! No! No! We say this three times to emphasize that after a vasectomy there are some active sperm located above the area of the where the vas is divided. It may take a dozen to two dozen ejaculations to clear the sperm from where the vasectomy is performed. You and your partner should use other forms of birth control until the doctor has had a chance to check your semen specimens by looking under the microscope on two separate occasions, to make sure the specimen is completely free of sperm.

Is the No-scalpel Vasectomy Safe?

Vasectomy in general is safe and simple. Vasectomy is an operation and all surgery has some risk such as bleeding, infection, and pain, but serious problems are unusual. There is always a small chance of the tubes rejoining themselves, and this is the reason that sperm checks are necessary. There have been some controversies in the past about the long-term effects of vasectomy, but to our knowledge there are no long-term risks to vasectomy.

How Long Will the No-scalpel Vasectomy Take?

It depends on the surgeon, but on average, the operation takes between fifteen to thirty minutes.

When Can I Start Having Sex Again?

As a rule, we suggest waiting a week before having intercourse. Remember, however, that the vasectomy only divides the vas and has no effect on the sperm that are already beyond that point. It is important not to have unprotected intercourse until the absence of sperm from the ejaculate has been confirmed with two (2) negative sperm checks two weeks apart. In other words, keep using your usual form of birth control until you have been given the "all clear" by your urologist.

Can I Reverse My Vasectomy if I Change My Mind?

The decision to opt for a vasectomy remains a highly personal one in which the potential risks and benefits must be considered, including the possibility that you may change your mind. Vasectomy reversal is possible but success is not guaranteed and depends largely on how long ago it was done. So it is much better to consider it a permanent procedure. Also, a vasectomy reversal can be very expensive. If you want a vasectomy but may still want kids, consider freezing sperm before the surgery. (Your doctor can usually help you locate a lab that freezes and stores sperm "for a rainy day!")

What is the Cost of a Vasectomy?

Costs range from several hundred to several thousand dollars. You should ask the doctor if the cost includes the initial consultation, the cost of the procedure, and also the post procedure semen examinations.

What happened to Paco?

Paco had counseling with a urologist and discussed having a vasectomy. He and his wife talked it over and decided to proceed with the vasectomy. Paco had a no-needle, no-scalpel vasectomy in the doctor's office. The procedure took fifteen minutes and he left the office, drove home, and used an ice pack on his scrotum until he went to sleep. He refrained from exertion and sexual intimacy for one week. When Paco brought two semen samples to the doctor's office to be examined under a microscope, there were no sperm found in either semen sample. He was declared sterile and he was free to engage in sexual intimacy without any need for contraception. His libido and erections were unchanged, and he and his partner were very happy campers. They told many of their friends about their positive experience of having a vasectomy.

Bottom Line

A vasectomy is one of the best ways to achieve contraception. It is nearly 100 percent successful and can be done in the doctor's office under a local anesthetic, without the use of a needle or a scalpel. The procedure is accomplished in just a few minutes and most men have no pain or discomfort afterwards. This truly is the "prime cut!"

CHAPTER 7

Battling Low Testosterone—When the Grapes Turn to Raisins

Patrick deals with the same stresses that most men in their mid-fifties experience—long work hours, financial strain, and a little weight gain. Patrick attributed his ongoing fatigue to these factors. However, his wife, Nancy, noticed over the last year that Patrick would spend most of his evenings and weekends on the couch. He no longer had the energy for golf and other exertional activities. His wife also noticed that his mood seemed fairly unpredictable. When Patrick became increasingly inattentive in the bedroom, Nancy knew something was wrong. She convinced Patrick to make that long overdue appointment with his primary care physician.

What Is Low T?

Low testosterone, commonly referred to as "low T," is a condition that has recently exploded in the public eye. The onslaught of self-diagnoses has been fueled by the plethora of new testosterone replacement products and the vague definition of this ailment. In fact, most of the symptoms of low testosterone are also shared with other more common medical conditions. For instance, a low energy level has dozens of explanations, with low testosterone being one of the less likely causes. Nevertheless, testosterone blood levels begin to decrease by 1 percent per year starting at age thirty. By age sixty, more than one in five men will have low testosterone levels.[1]

Although there is no one governing organization that determines who is a good candidate for testosterone replacement, most reputable doctors would agree that replacement should only be considered in patients with *both* a low blood testosterone level *and* at least one symptom associated with low testosterone. For instance, if a man's serum testosterone level is normal, testosterone should not be administered to treat fatigue or sexual dysfunction. In any case, your physician should search for other causes of your symptoms before considering testosterone replacement especially if the testosterone level in the blood is normal.

To add to the confusion, there is also some disagreement on what blood levels are considered normal. The normal range of testosterone is 300–900 ng/dL. Typically, and any blood level below 300 ng/DL is considered low. However, this level can vary based on other factors such as time of day that the blood is drawn. It is generally recommended that the blood test be obtained in the morning when testosterone levels in the blood

1. Brawer, Michael K. "Testosterone Replacement in Men with Andropause: An Overview." *Reviews in Urology.* 2004. Accessed August 01, 2017. https://www.ncbi.nlm.nih.gov/pmc/articles/PMC1472881/.

are at a peak. If a low level is obtained late in the day, the test should be repeated in the morning. Repeat testing is not needed if the afternoon blood draw is in the normal range.

For borderline testosterone levels, additional testing should be performed. The standard test that we have been discussing is commonly referred to as a "total testosterone" level. This measures both free-floating testosterone and testosterone bound to proteins in the blood stream. It is the "free" testosterone that is the active component and the hormone that is responsible for many functions including libido, bone strength, muscle mass, energy level, and even a man's mood. As such, the typical symptoms associated with low testosterone include low sex drive (libido), decreased erectile function, loss of muscle mass, decreased bone mass (osteoporosis), mood changes, depression, increased body fat, and fatigue. Although there are ways to either calculate or directly measure free testosterone levels, this test to measure free testosterone is expensive and has fallen mostly out of favor. The best test for the borderline patients is "bio-available testosterone."

So back to the question, what is low T? Some would say it is merely a low testosterone level (or bio-available level in the borderline patient). Others would say that symptoms usually associated with low testosterone levels meet the criteria, even if the blood levels are not truly low (or even worse, never measured). The name "low T" is actually a popular renaming of the well-established medical condition, hypogonadism. It is not just a chemical (blood test) diagnosis, nor is the term just a clinical diagnosis. Hypogonadism (also called androgen deficiency) is the condition in which the testicles do not produce the proper amount of testosterone *and*, as a result, clinical symptoms consistent with testosterone deficiency develop.

Causes of Low T

The most common cause of androgen deficiency is low production of luteinizing hormone (LH) by the pituitary gland. In fact, when testosterone levels are low, the pituitary gland should actually produce high levels of LH in an attempt to stimulate the testicles to produce more testosterone. The decrease in LH production in the middle-age and older men occurs for unknown

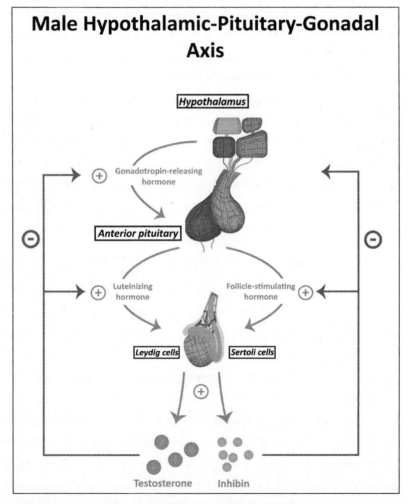

Figure 1: The Pit Bosses, the Balls *(Shutterstock)*

reasons. With the decrease in LH from the pituitary gland, the testicles are not stimulated to produce testosterone. This condition is called secondary hypogonadism. On rare occasions, this can be caused by a benign tumor of the pituitary gland that can only be diagnosed with a special imaging scan, the magnetic resonance imaging or MRI, when suspected.

Primary hypogonadism occurs when the testicles are damaged and unable to produce adequate amounts of testosterone. Examples of primary hypogonadism include undescended testicle (see Chapter 8), prior testicular infections, prior chemotherapy, and previous inguinal or scrotal surgery.

Other causes of low testosterone disturb the interaction between the pituitary gland and the testicles. These include diabetes, chronic renal disease, sleep apnea, use of opiates, use of certain antidepressants (Celexa®, Lexapro®, Prozac®, Paxil®, Zoloft®), drugs used to treat prostate cancer, chronic liver failure, alcoholism, drug abuse, and depression.

The Risks of Testosterone Replacement

So what would be the problem with just trying testosterone replacement and seeing if it works? Some men with "normal" testosterone levels may feel that their level might not be normal for them. The first problem with that approach is that a man with the right amount of testosterone can feel better with an excess amount. In fact, for this reason, testosterone can be very addictive. One fairly common example is the bodybuilder that uses excess doses of testosterone to build muscle without any medical oversight. It is this addictive nature of the drug that caused the FDA to regulate testosterone as a controlled substance, similar to that of narcotic drugs. As such, the FDA does not permit prescribing testosterone over the phone and a written prescription must be obtained from the doctor in most states.

Testosterone replacement can also affect how your body's own natural testosterone is produced. The pituitary gland is a small gland located below the brain that regulates how certain hormones in the body are produced. If this little gland detects testosterone in the blood from *any* source, it will tell the testicles to stop producing testosterone. With this concept in mind, testosterone replacement is just that—"replacement." Otherwise, it would have been called testosterone "enhancement." The longer a man receives testosterone replacement therapy, the longer it will take for his natural production to return once the replacement is stopped.

The pituitary gland also affects sperm production in the testicle. When receiving testosterone, the pituitary gland will stop sending these hormonal signals to the testicle. As a result, production of sperm and the quality of the sperm will plummet. This effect on fertility is usually reversible. However, for men who are interested in maintaining their fertility, a drug that is similar to the pituitary hormone (oral Clomid or subcutaneous HCG) is given instead of testosterone. Clomid and HCG will cause the man's own testicles to produce more testosterone, yet not adversely affect sperm production.

Too much testosterone can cause overproduction of red blood cells. This is an example of "too much of a good thing." If the red blood cell count or hematocrit (percentage of blood that is made of cells versus liquid) rises much above 50 percent, our blood becomes too thick. This excess blood may clog small arteries—such as those going to the heart or brain—thereby increasing the risk of heart attack or stroke. This condition is different than blood clot formation which is a biochemical process. Having too many red blood cells is simply a mechanical problem that is not affected by aspirin or other blood thinners. Most men with too many red blood cells can be easily managed by having the man donate blood at a blood bank and having his blood count tested a few weeks later. (It is of interest that most blood banks will take the blood from men using testosterone but not transfuse it into other patients in need of blood.)

Testosterone also affects the prostate. It will potentially increase the prostate size, disturb urinary habits, and raise the PSA levels (blood test for prostate cancer screening, see Chapter 3). Although there is no evidence that testosterone causes prostate cancer, careful monitoring of the PSA level and having a digital examination is recommended during testosterone replacement. Most doctors will recommend that men receiving testosterone replacement should have the PSA and digital rectal exam every six to twelve months.

Several reports have shown that testosterone replacement increases the risk of having a cardiovascular event (even with normal red blood cell count).[2] However, these studies involve a population of men with other significant cardiovascular risk factors. Careful risk assessment and monitoring should be conducted on any man considered for testosterone replacement. In men who are at a higher risk for cardiovascular disease, replacement should be avoided despite any androgen deficiency symptoms or low blood testosterone levels.

---

2. G, Giovanni Corona, Giulia Rastrelli, Elisa Maseroli, Alessandra Sforza, and Mario Maggi. "Testosterone Replacement Therapy and Cardiovascular Risk: A Review." *The World Journal of Men's Health*. December 2015. Accessed August 01, 2017. https://www.ncbi .nlm.nih.gov/pmc/articles/PMC4709429/.

Monitoring should include regular blood counts, testosterone levels, PSA blood test, and liver function tests. Men should only receive the minimum amount of testosterone necessary to achieve the essential goals.

Risks of Low Testosterone

Testosterone is an essential hormone. It is partly responsible for maintaining lean body mass, bone health, and red blood cell production. Testosterone deficiency increases the risk of obesity, diabetes, metabolic syndrome, osteoporosis, and clinical depression. Interestingly, weight loss, diabetes control, moderate exercise (especially weight-lifting), better sleeping habits, and even sex can significantly improve testosterone levels. This is a good example of "the chicken and the egg." Is the deficiency causing the symptoms, or are the symptoms causing the deficiency? It is our opinion that it is probably both.

A Special Note on Cardiovascular Effects

Testosterone replacement in certain patients can increase the risk of cardiovascular events such as stroke and heart attack. However, we have long known that men with androgen *deficiency* are also more likely to experience a cardiovascular event.[3] So which one is true? Probably both.

So, what are the explanations behind the seemingly contradictory ends of the spectrum. With testosterone deficiency, for

---

3. Morris, Paul D., and Kevin S. Channer. "Testosterone and cardiovascular disease in men." *Asian Journal of Andrology*. May 2012. Accessed August 01, 2017. https://www.ncbi.nlm.nih.gov/pmc/articles/PMC3720171/.

example, obesity is more common but certainly not heart-healthy. However, patients with existing heart disease are more sensitive to changes in blood volume. Testosterone replacement can cause some fluid retention in these patients and put further stress on the heart. Every man has an ideal range for their testosterone levels. Too much testosterone is not good; too little is not good either. This concept applies to many of our bodily demands. Just like little Goldilocks in the cottage of the three bears: one bowl is too hot, another is too cold, but the third one is just right. The goal of your doctor is to help you get to the testosterone level that is just right for you in order to improve your symptoms of low T.

A Special Note on Prostate Cancer Patients and Those at Risk for Prostate Cancer

Testosterone replacement has never been shown to cause prostate cancer. However, special care and monitoring should be given to those men who are at increased risk for developing prostate cancer (African Americans, men with a family history of prostate cancer, men with elevated PSA blood tests, or men with an abnormal digital rectal exam). However, the same rules apply for these men wishing to have testosterone replacement therapy: low testosterone blood level, significant symptoms consistent with androgen deficiency, and minimum dosing to achieve essential therapeutic goals.

For men who have been treated for prostate cancer, a urologist's input is essential. For low to medium risk prostate cancer patients who have seemingly been successfully treated, testosterone replacement is reasonable. The goal is to put men with androgen deficiency on the same playing field (with the same energy level and same sex drive) as the average guy with a normal T level, albeit with careful dosing and monitoring. After all, for the average guy with normal testosterone production, we would not put them on medications to lower their testosterone. As such, we should not deny treatment for low T in men who are seemingly cured of their prostate cancer. It is very important for men who have been treated for prostate cancer either with radiation therapy or surgical removal of the prostate gland have a PSA level that has remained very low for several months prior to initiating testosterone replacement. These men must commit to regular PSA testing. If the PSA increases following testosterone replacement therapy, the treatment for low T must be stopped immediately. Even in the worst case scenario, in a patient who was not cured of their prostate cancer, the PSA will eventually begin to rise and reveal the residual disease. The testosterone replacement would then cease, and additional treatment for the prostate cancer would be initiated

Initiating Treatment

Your physician will first conduct a detailed history and physical exam in order to confirm the diagnosis as described above and also possibly to determine the underlying cause. Repeat blood tests are often necessary. These tests would include a testosterone level, a PSA level, and pituitary hormones in some cases. Your medical history will also help assess your risks associated with hormone deficiency and replacement.

Methods of Administration

Oral medications are usually not an option. Testosterone pills have a high likelihood of damaging the liver compared to other routes of administration. Despite the obvious convenience, because of these side effects, oral testosterone is a poor choice.

Topical dosing through the skin comes in three forms: patch, gel, and underarm liquid. These topical treatments all require daily morning application and probably mimic natural testosterone production better than other methods of replacement. However, some men find these topical testosterones inconvenient, messy, and easy to forget to use every day. Care must be taken not to transfer the drug to the skin of another person (especially children). It is recommended that all men carefully wash their hands after the application of the testosterone gel. Cost can also be very high, although this can often be remedied in part by using a compounding pharmacy.

Intramuscular injections of testosterone require an injection with a needle every one to two weeks. This treatment option is the most common method of testosterone delivery, probably since it has been around the longest. However, it is very difficult to maintain consistent levels. For instance, during the first few days following the injection, the levels can be very high and even cause euphoria. Conversely, in the days leading up to the next injection, the level can fall precipitously and cause a major downswing in mood and energy. We often refer to this as the "yo-yo effect."

A somewhat novel method uses a very small wafer placed at the gum line on the inner cheek. This is called a buccal patch. However, most men do not like keeping something in their mouth for long periods of time. Perhaps the exception is a pinch of Skoal!

A longer lasting option involves implantation of testosterone pellets below the skin usually in the area of the buttocks. These small rice-sized pellets typically last six months. However, as with the injections, the initial levels can be too high, and the later levels (the last month) can be too low. It is also difficult to reverse, as it is almost impossible to remove the pellets once they have been inserted under the skin. Since testosterone pellets are difficult or impossible to remove, they should not be used in men who have been treated for prostate cancer since it would be difficult to remove the small pellets if the PSA were to rise.

Testosterone patches are called transdermal patches and they are applied to the skin. The testosterone in the patch, like the topical gels, will pass through the skin and enter the blood stream where it can supply the organs and body parts that are in need of testosterone. Patches are administered and replaced once a day. The patches are worn on the back, upper arm, thigh, or abdomen. One brand of testosterone patch is applied to the scrotum. It may take several weeks before men will start to feel the beneficial effects of the testosterone patch. The most common problem with these patches is skin irritation with visible red blotches that are often itchy or uncomfortable.

Side Effects of Testosterone Treatment

As mentioned earlier, testosterone therapy can raise a man's risk for blood clots and stroke. This is a result of the stimulation of the red blood cell production. We recommend that all men on testosterone replacement therapy have a complete blood count (or CBC) blood test every six months. If the red blood cell count is significantly increased, then men will be advised to donate blood at a blood bank and have the blood count repeated in a few weeks to see that the count is in the normal range. At times, the testosterone dosing can be lowered to achieve an acceptable red blood cell count. However, in some men, regular blood donation will become a part of the treatment.

Uncommon side effects of testosterone replacement include sleep apnea, acne, and breast enlargement (or gynecomastia). All of these side effects are transient and will go away if testosterone treatment is discontinued.

Men who use a testosterone gel should wash their hands thoroughly after applying topical testosterone and make certain that no one else touches the spots where they apply the medication. If a woman or child comes into contact with testosterone gels, it can cause side effects in them, including hair growth in women and premature puberty in young children.

Contraindications for Using Testosterone Replacement Therapy

Testosterone replacement is contraindicated in men who have breast cancer or untreated prostate cancer, high red blood cell count, congestive heart failure (CHF), liver disease, and significant water retention. Caution should also be taken in men with sleep apnea, benign prostatic enlargement, and cardiovascular disease.

The Benefits of Testosterone Replacement

Despite the potential perils, testosterone is an essential hormone. Return to a normal level can lead to the following benefits: enhanced libido, increased muscle mass, decreased fat mass, better control of diabetes, correction of anemia, improvement in bone strength, mood improvement, sense of well-being, and improvement in cholesterol and lipid levels.

"Andropause"—the Pause That Hardly Refreshes

It's interesting that the word menopause contains the word "men." Well, andropause is the male equivalent of menopause. Like many topics in this chapter, the concept of andropause is controversial. Testosterone levels naturally decline with each decade of life. Some argue that this is a normal process that should not be treated. However, there are no well-established blood testosterone levels that have been adjusted for age.

The Androgen Deficiency in the Aging Male (ADAM) questionnaire[4] about symptoms of low testosterone was developed to help men describe the kind and severity of their low testosterone symptoms. Here are the ten simple questions:

1. Do you have a decrease in libido (sex drive)?
2. Do you have a lack of energy?
3. Do you have a decrease in strength and/or endurance?
4. Have you lost height?
5. Have you noticed a decreased "enjoyment of life?"
6. Are you sad and/or grumpy?
7. Are your erections less strong?
8. Have you noticed a recent deterioration in your ability to play sports?
9. Are you falling asleep after dinner?
10. Has there been a recent deterioration in your work performance?

---

4. Tancredi, A., JY Reginster, F. Schleich, G. Pire, P. Maassen, and F. Luyckx And. "A Tancredi." *European Journal of Endocrinology.* September 01, 2004. Accessed August 01, 2017. http://www.eje-online.org/content/151/3/355.

If you answer "yes" to number 1 or 7 or if you answer "yes" to more than three questions, you may have low testosterone.

Like most things in medicine, treatments should be individualized. Testosterone replacement should not be considered as the fountain of youth. As much as every man would like, testosterone replacement will not turn back the clock. Administration should be targeted towards reduction of the harmful health effects from declining testosterone production.

So What Happened to Patrick?

After a complete evaluation, his physician did determine that he meets all the criteria for clinical androgen deficiency. He started on a daily testosterone gel preparation and his testosterone blood level went from 150 dL/mL to 450 dL/mL. Although not perfect, his mood, energy level, and libido dramatically improved. Nancy does not describe him as "a new man." She describes him as "the man she always knew." And Pat feels like a real Patrick, not a Patricia.

Bottom Line

Testosterone deficiency affects millions of middle-age and older men. Many of men with low T levels will have symptoms such as low energy levels, decrease in muscle mass, decrease in bone mineral density, and a negative effect on a man's mood. The diagnosis of low T is easily made with a blood test best taken in the morning. Most men with low T and symptoms of low T are candidates for testosterone replacement therapy. There are multiple options for treating low T including injections, topical testosterone, and testosterone pellets. Men receiving testosterone replacement should have regular checkups for their PSA levels and their red blood cell counts. So if you aren't feeling like as much of a man as you would want, then speak to your doctor and have a testosterone blood test.

CHAPTER 8

Problems in the Pouch— When the Family Jewels Don't Shine

Pomeroy's wife was flipping through her husband's recent issue of Men's Health Magazine, *when she noticed an article about men checking their bags. And no, she was not reading the travel section. Of course, she left it open to that page and left it on the kitchen counter for her Pomeroy to read. The next morning, Pomeroy discovered by self-examination a small and tender lump "down there." Pomeroy could not make an appointment with the urologist quickly enough.*

Scrotal Anatomy: How Is That Bag Packed?

Most men do not pay attention to what's going on down there until there is a problem. However, some problems down there do not cause any symptoms. It is also important to be familiar with your particular anatomy in order to recognize any abnormal changes.[1]

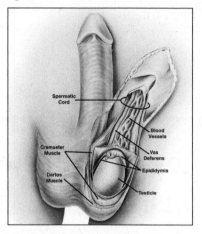

Figure 1: The Scrotum *(iStock)*

The skin of the scrotum is very specialized. As you probably know, the testicles are located outside of the body in order to keep the family jewels at the proper temperature. Normal body temperature is a bit too balmy for those freshly made sperm. The scrotal skin contains the dartos muscle that raises and lowers the testicles closer and further from the body in order to regulate their finicky temperature needs. During this process, the skin will wrinkle, so don't be tempted to iron it (even if you have a date).

The scrotum contains a lot more than two balls. Behind each testicle is a structure called the epididymis. It is a smooth but soft coil of small tubes that store sperm previously manufactured in the testicles. The sperm exit the epididymis through a tube called the vas deferens (or "vas" as in "vasectomy"). The vas travels into the abdomen through the groin (also called

---

1. Campbell, Meredith F., Alan J. Wein, Louis R. Kavoussi, Alan W. Partin, Craig A. Peters, and Patrick C. Walsh. *Campbell-Walsh Urology*. Philadelphia: Elsevier, 2016.

"inguinal canal") alongside the vas are the testicular artery and veins. Collectively, the vas tube and these blood vessels are referred to as the spermatic cord. This spermatic cord also contains the cremaster muscle which aids the dartos muscle in regulating the scrotal temperature and keep the testicles a degree or two cooler than then core temperature of the body of 98.6°F. The cremaster also contracts in response to any physical threat to the scrotum such as a swift kick, causing the testicles to rise and stay out of harm's way.

The testicles basically perform two functions: sperm production and testosterone production. Anything that can damage the testicles can also affect these two functions. Both of these functions are regulated by hormones produced by the pituitary gland (a small gland located just below the brain).

A Problem in the Sack

No, we're not talking about poor performance in the bedroom. However, problems in the scrotum can lead to issues with sexual prowess and fertility. We will spend the remainder of this chapter discussing such problems as pain, abnormal swelling, and even cancer. And, of course, we'll delve into the solutions for these problems.

An Ouch in the Pouch—Scrotal Pain

The two most common causes of scrotal pain are infection and trauma. On rare occasions, prior vasectomy or hernia repair can cause chronic pain. However, sometimes the cause is not so obvious. In some cases, ball pain can become a long-term and difficult-to-treat problem.

The first step in evaluating testicular pain is differentiating whether it is tender or just painful. If it is tender, the problem usually originates in the testicle or epididymis (such as an infection, covered later in this chapter). However, if it is painful but not tender to the touch, the pain could be coming from somewhere else. This condition is called referred pain. A common example of referred pain occurs during a heart attack, during which the pain "radiates" to the left arm. Of course, the heart is not located in the arm. Similarly, the painful passage of a kidney stone can cause pain in the testicle. This phenomenon occurs because the sensory nerves of the kidney enter the spinal cord in a similar location to the testicular nerves. Other examples of referred testicular pain include inguinal hernia, prior hernia repair, and prostate problems.

A careful history and physical exam can often determine the underlying cause. In other cases, an ultrasound or CT scan may be helpful. When the cause cannot be either determined or treated, therapy is directed towards the pain itself. Anti-inflammatory, anti-spasmodic, and anti-seizure (for example, Neurontin®) medications have all been tried with varying successes. A pain specialist physician may play a role with such treatments as a nerve block. This procedure is initially performed with a local anesthetic injection in the nerves leading to the testicle. If this temporary treatment is successful, a longer-lasting destruction of the nerve could be performed using radiofrequency (heat) energy.

A relatively new option for controlling chronic pain is implantation of an electrical nerve stimulator (Interstim™ device). As a last resort, the nerves leading to the testicle can be surgically cut. Although this procedure can be performed with a laparoscopic technique, the nerves must be transected in the back portion of the abdomen. With any of these procedures, it is not unusual for the pain to return within one to two years. In any case, a man with such a severe case of testicular pain must be his own advocate and seek advice from multiple specialists until he finds an answer. In its extreme form, it can be very difficult for man and his loved one to cope with this condition.

Bulging in a Bad Way: The Groin Hernia

The body is made of many compartments, and every structure is assigned to its own compartment. When a part of our anatomy either can or does go from its own compartment into another compartment, it is called a hernia. For instance, if the stomach can protrude through the diaphragm and into the chest, it is called a hiatal hernia. The most common type of hernia is an inguinal, or groin, hernia. In this case, an abdominal structure such as the intestine can protrude from the abdomen into the scrotum or into the area just above the scrotum (the inguinal canal).

The most common symptom from an inguinal hernia is testicular or groin pain. Often, there are no symptoms, and a man or his physician merely notices a bulge in the area. In most cases, a groin hernia requires surgical correction. If left untreated, a portion of the intestine can become trapped outside of the abdomen leading to a surgical emergency.

A hernia repair is typically a safe, outpatient procedure. A man can usually return to work in a couple of days following the procedure, and he could typically resume a light exercise routine within one to two weeks. For most hernia surgeries, a piece of permanent synthetic mesh is used for the reconstruction. Problems with the use of mesh are rare, but there are reported cases of entrapment of the testicular nerve causing long-term pain. This complication would possibly require removal of the mesh in order to release the nerve.

Extra Parts—Cysts, Hydrocele, Varicocele

Some people are blessed with bigger parts, and some are blessed with extra parts. Cysts are fairly common in the scrotum. A cyst is merely a thin-walled sphere containing fluid. They come in a dramatic variation of sizes. When located in the epididymis, adjacent to the testicle, they are often called spermatoceles since they contain sperm. No treatment is required unless they cause pain.

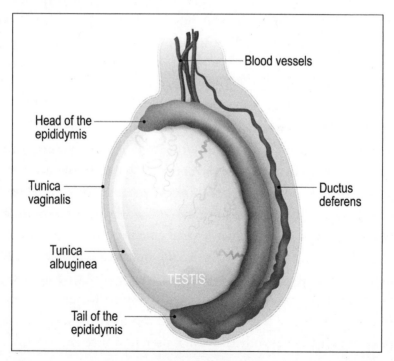

Figure 2: Anatomy of the Testicle *(iStock)*

Surrounding the testicle is a thin sac that contains a small amount of fluid which lubricates the testicle while it bounces around. If this fluid builds up, it becomes a hydrocele. Like the spermatocele, no treatment is required unless it causes pain. However, if a hydrocele contains so much fluid that the testicle cannot be felt, it should be evaluated by a urologist.

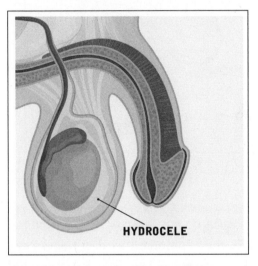

HYDROCELE

Figure 3: Hydrocele *(Shutterstock)*

If you feel a "bag of worms" down there, you may have a varicocele. A varicocele is the testicular version of varicose veins that are common in the legs. With this condition, the little valves within the veins do not work properly, and the blood backs up. As a result, the veins can become very dilated. Interestingly, when you lie down, the blood will drain better and the veins will become smaller. However, you should still see your doctor to make the final diagnosis. Because of our internal anatomy, varicoceles are much more common on the left. In fact, if it is only present or more predominant on the right, an abdominal tumor could be an underlying cause.

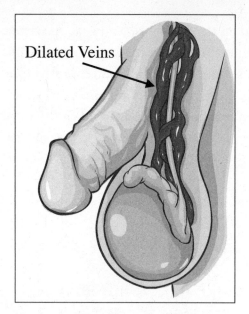

Dilated Veins

Figure 4: Varicocele *(iStock)*

Most varicoceles do not cause symptoms, but some men will have pain. Also, the additional blood in the scrotum raises the temperature down there and could cause fertility problems. The fix is a minor surgical procedure in which the urologist ties off the veins through a small incision above the scrotum.

Your physician can usually diagnose those extra parts down there with a simple physical exam. A scrotal ultrasound can confirm the diagnosis.

Taking a Blow Down There: Testicular Trauma

Most men have experienced that very distinct pain associated with a foot or projectile to the crouch. Since the nerves from the testicles enter the spinal cord in a nearby location to the intestines, severe nausea can quickly follow trauma to the jewels. In some cases, the pain can be so intense that it causes a man to black out.

Most injuries resolve on their own with nothing more than a little tender loving care. However, if the testicle were to rupture, it becomes a surgical emergency. Any significant swelling or bruising should raise this concern. The diagnosis is easily made with an ultrasound of the scrotum. Failure to repair a ruptured testicle often leads to loss of that precious jewel.

Serious injuries to the testicle can also damage other parts. Any scrotal skin lacerations should be promptly treated. The urethra (the tube through which urine exits the bladder) passes just behind the scrotum prior to traveling though the penis. If the testicle is injured by a "straddle" injury, the urethra can sustain significant damage. Blood at the tip of the penis or in the urine is one of the indicators of a urethral injury. Sometimes, a man will not notice the signs of urethral trauma in the face of severe testicular pain. As always, when in doubt, check it out.

The Skinny on Scrotal Skin

Almost anything that can affect the skin elsewhere can affect the scrotal skin. Because of the predominance of hair follicles and sweat glands in a bacteria-rich environment, sebaceous cysts are very common down there. These are round nodules contained in the scrotal wall, separate from any of the internal structures described earlier. Usually, these should be removed to prevent any future problems.

Jock itch is also a common scrotal skin problem. The red rash and itching can sometimes be painful. This fungal infection tends to affect the moister area between the scrotum and the thigh. Careful hygiene and an antifungal cream or spray powder usually eliminates the problem.

Warning: If you notice any infected or tender pimple or boil on (or near) the scrotum, seek medical attention immediately. These infections can spread rapidly, destroy large amounts of skin and muscle, and become life-threatening—especially in diabetics or immunocompromised men. For those who chose to remove hair down there, clippers are far superior to razors to prevent ingrown hairs and subsequent skin infections.

Well, That's Just Swell—Scrotal Edema

The scrotum can often serve as a pop-off valve for the body. Any fluid build-up in the area can easily cause the scrotum to expand to several times its normal size. Any fluid or blood from abdominal surgery will easily travel to the scrotum.

The good news is that the scrotum will return to normal with time. Elevating your swollen sack with a folded hand towel may help reduce the size. The swelling will not restrict urine flow, but it can make your aim even worse.

Infection in a Bad Direction

Bacteria from the genitourinary or gastrointestinal tracts can make their way into the scrotal sack and cause an infection in the epididymis (epididymitis), testicle (orchitis), or both (epidid-ymo-orchitis). The testicle can become very painful, hard, and swollen. In more severe cases, bacteria can spread to the blood stream and cause high fevers. The diagnosis is usually clinical, but it can be confirmed with a scrotal ultrasound. It is important to distinguish this diagnosis from testicle torsion (discussed below).

Some cases require two to three weeks of antibiotics, yet the symptoms may take over a month to resolve. Nonsteroidal anti-inflammatory medications can help relieve the symptoms and hasten the recovery. If the infection is also in the urine, it can be tested to confirm antibiotic effectiveness. Otherwise, clinical judgement is based on age and sexual exposure (discussed further in Chapter 12).

Now That's Twisted—Testicular Torsion

Some testicles can be a little more mobile than others. If one becomes twisted, it can choke off the blood supply and cause the testicle to die. This is a true surgical emergency requiring the urologist to surgically untwist the spermatic cord and suture the testicle to the inside of the scrotal wall in order to prevent repeat twisting. The other (seemingly normal) testicle would also be sutured in place to prevent it from twisting in the future.

This condition is most common during the early teen years and is very rare much beyond age thirty.[2] Testicular torsion should be a consideration in any man with sudden pain and swelling of the testicle. The diagnosis is easily confirmed or eliminated with an ultrasound that can detect the presence or absence of blood flow.

On occasion, testicular torsion can be intermittent. In these cases, the diagnosis is difficult to prove. If the story is

2. "Medical Student Curriculum: Acute Scrotum." American Urological Association—Medical Student Curriculum: Acute Scrotum. Accessed August 01, 2017. http://www.auanet.org/education/auauniversity /medical-student-education/medical-student-curriculum/ acute-scrotum.

convincing, both testicles should be sutured to the inside layer of the scrotal skin to prevent future torsion.

Recovery from this surgery is fairly quick. If performed within four hours of onset, very little damage will occur. Within six to eight weeks, you will not be able to notice any effects from the surgery. If left untreated, removal of the testicle is often the best option.

Undescended Testicles

Undescended testicles, also called cryptorchidism, occur when the testicle fails to fall from the abdomen to the scrotum while in utero or during the first year of life. Any man with a history of an undescended testicle is at higher risk for cancer in either testicle—even the normal one.[3] Therefore, regular testicular self-exam becomes even more important. Cryptorchidism also increases the risk of future fertility issues.

An undescended testicle is usually addressed during a child's first year of life. The testicle can be surgically pulled down into the scrotum. Although this surgery will potentially improve fertility, it will not reduce the already higher incidence of testicular cancer. When detected later in life, removal is usually best since the function becomes minimal and the risk of cancer is higher.

---

3. Akre, O., A. Pettersson, and L. Richiardi. "Risk of contralateral testicular cancer among men with unilaterally undescended testis: a meta analysis." *International Journal of Cancer.* February 01, 2009. Accessed September 01, 2017. https://www.ncbi.nlm.nih.gov/pubmed/18973229.

Ugh, The "C" Word

Despite the public awareness following Lance Armstrong's diagnosis, testicular cancer is relatively rare. Regular testicular self-exams are recommended for early detection. It is characterized by a painless lump (unlike torsion or orchitis) on or in the testicle. If it is separately moveable from the testicle—such as being attached to the epididymis—it is most likely not a cancer. The diagnosis can be confirmed with a scrotal ultrasound.

Testicular self-exam should be performed monthly while taking a warm shower so that the scrotal skin is loose and thin. Systematically allow *one* of the testicles to roll through your fingers using both hands for the best control. Learn which

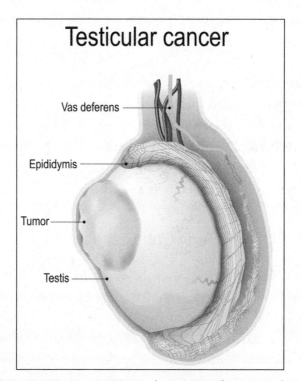

Figure 5: Worrisome Testicular Mass *(Shutterstock)*

part of the scrotal contents is the epididymis so that you can recognize what belongs. Perform the same on the other side. A lump or hard spot that is part of the testicle should be brought to attention of a urologist. If the lump is clearly separate from the testicle—such as part of the epididymis—it may be safe for you just to monitor it. When in doubt, have the doctor check it out.

The initial treatment for a suspicious testicular mass is removal through a groin (lower abdominal) incision. The surgical incision is never performed on the scrotum since this could cause unpredictable spread of the cancer. Some special testicular cancer blood tests will be drawn prior to the surgery (alpha-fetoprotein, AFP; beta human chorionic gonadotropin, ß-HCG; and lactate dehydrogenase, LDH). A CT scan of the chest, abdomen, and pelvis will help determine if there is any spread requiring chemotherapy, radiation, or further surgery. Sometimes in the absence of spread, these treatments are offered as preventive measures in higher risk patients.

Even when testicular cancer spreads to distant areas in the body, the cure rates with chemotherapy are in the high nineties. However, catching it early on self-exam will minimize the necessary treatment and maximize the cure.

Boxers Versus Briefs

No scrotal chapter would be complete without discussing the never-ending controversy of boxers versus briefs. The answer is that there is no answer. However, the debate remains as heated as some across-the-aisle political discussions. Briefs tend to keep the junk supported and out of harm's way. Boxers are more breathable. Boxers allow the scrotum to regulate its own temperature, which some say may lead to healthier sperm counts. However, neither style influences testosterone production or sexual prowess—unless your partner prefers one look over the other. Did I hear commando?

So What Happened to Pomeroy?

Pomeroy saw the urologist who diagnosed him with a spermatocele. The doctor did not think there was any infection and just prescribed over-the-counter anti-inflammatory medication. His symptoms resolved in a few days, although the cyst has remained unchanged in size for over one year. Most importantly, this "extra part" was clearly separate from the testicle, essentially eliminating the possibility of cancer. Nevertheless, Pomeroy continues to perform regular testicular self-exams.

Bottom Line

There's a lot of stuff hanging around down there, and a lot can go wrong. If something doesn't feel or look right, see your doctor right away. Whether it's a source of pain or a possible cancer, waiting too long can make it much more difficult to treat. Unlike at the airport, there is no cost to checking your bag.

CHAPTER 9

Osteoporosis in Men—We Don't Like Soft Bones

contributed by Mindi S. Miller, PharmD, BCPS

Jeff is a sixty-three-year-old truck driver who smokes cigarettes and enjoys a few beers two to three times a week. Recently, Jeff had a minor fall after slipping on a patch of ice. An x-ray revealed that the fall caused a hip fracture. The emergency room physician was concerned, so he sent Jeff for a bone density test. Jeff was shocked to find out that he has osteoporosis. He was under the impression that osteoporosis is a condition that only occurs in women.

Are Men Getting Soft?

Osteoporosis is a thinning or softening of the bones. The word "osteoporosis" literally means "porous bones," and is often considered a women's disease because it is more common in women and tends to occur at a younger age in females.

However, we now recognize that osteoporosis is a serious threat for men as well.

More than eight million men in the U.S. have osteoporosis or low bone mass.[1] Also, 20 percent of men over the age of fifty will suffer from a fracture that is related to osteoporosis. What's more, men are two to three times more likely to die after a hip fracture than their female counterparts.[2]

Osteoporosis is often called a silent epidemic because there are no symptoms. Most people don't realize that they suffer from the disorder until a fracture occurs.

The Blame Game

Bone is made up of collagen, a fibrous protein that forms the structure of bone, along with minerals such as calcium. Our bones are living tissues that are constantly changing. The body continuously removes old bone and replaces it with new bone material. During childhood and our early adult years, the skeleton grows in size and strength. The amount of new bone produced by men is significantly more than produced by women. Adults reach their maximum bone density between the ages of twenty and thirty.[3] After we reach the peak bone mass, the removal of bone exceeds the formation. As a result, bones

1. Finkelstein, Joel S., MD. "Treatment of Osteoporosis in Men." UpToDate. July 25, 2016. Accessed January 01, 2017. http://www .uptodate.com/.
2. "International Osteoporosis Foundation | Bone Health." International Osteoporosis Foundation | Bone Health. October 01, 2014. Accessed January 10, 2017. https://www.iofbonehealth.org/.
3. Ibid.

slowly become thinner, weaker, and more likely to break. If the bones lose enough mass, the patient develops osteoporosis. Certain risks can speed up the loss of bone mass.

Risky Business—Don't Be a Bonehead

There are two types of osteoporosis: primary and secondary. Primary osteoporosis is caused by bone loss related to aging or to an unknown cause. Most men with osteoporosis have at least one risk factor or secondary cause. Risk factors for osteoporosis are listed in Table 1. The most common causes of secondary osteoporosis in men include hypogonadism (low testosterone levels), alcohol abuse, smoking, use of glucocorticoid and other medications (Table 2), gastrointestinal disorders and poor nutrition, immobilization, and too much calcium in the urine (hypercalcemia).

Table 1
Risk Factors for Osteoporosis in Men

Alcohol abuse	Chronic	HIV
Physical inactivity	obstructive	Celiac disease
Low body weight	pulmonary	Parathyroid
Calcium or	disease	disease
vitamin D	Asthma	Thyroid disease
deficiency	Cystic Fibrosis	Inflammatory
Smoking	Hypercalciuria	bowel disease
Low testosterone	Rheumatoid	(Crohn's
levels	arthritis	Disease,
Medications	Chronic kidney or	Ulcerative
(see table 2)	liver disease	Colitis)
Androgen	Diabetes Mellitus	Transplants
deprivation	Cushing's	Cancer
therapy (for	syndrome	
prostate cancer)	Family history	

Table 2
Medications that can cause osteoporosis

Glucocorticoids (used for inflammation)
Prednisone
Hydrocortisone
Methylprednisolone
Others
Anticonvulsants (used for seizures)
Phenytoin (Dilantin)
Phenobarbital
Carbamazepine (Tegretol and other brand names)
Primidone (Mysoline)
Anticoagulants (Blood thinners)
Heparin
Antidepressants/mood stabilizers
Fluoxetine (Prozac)
Paroxetine (Paxil)
Sertraline (Zoloft)
Lithium
Others
Cancer Chemotherapy Drugs
Anastrozole (Arimidex)
Exemestane (Aromasin)
Letrozole (Femara)
Methotrexate
Diabetes drugs known as thiazolidinediones (TZDs)

Rosiglitazone (Avandia)
Pioglitazone (Actos)
Proton Pump Inhibitors (used for acid reflux; heartburn)
Omeprazole (Prilosec)
Pantoprazole (Protonix)
Esomeprazole (Nexium)
Lansoprazole (Prevacid)
Loop Diuretics (used for high blood pressure, heart failure)
Furosemide (Lasix)
Torsemide (Demadex)
Bumetanide (Bumex)
Thyroid hormones in excess
Levothyroxine (Synthroid)
Armour Thyroid
Aluminum-Containing Antacids
Gaviscon
Maalox
Mylanta
Immunosuppressants (used for transplant patients to prevent rejection)
Cyclosporine A (Sandimmune and other brand names)
Tacrolimus (Prograf and other brand names)

Hormone Levels

Abnormally low levels of sex hormones, known as hypogonadism, is a known cause of osteoporosis. Up to thirty percent of men who suffer vertebral (spinal) fractures due to osteoporosis have low testosterone levels.[4] It is normal for these levels to decrease as a man ages, but testosterone levels should not drop suddenly (as estrogen levels do in women at menopause). Some medications, such as antiandrogens used in the treatment of prostate cancer, can greatly reduce testosterone levels. Interestingly, the evidence shows that low levels of estrogen in men also contributes to osteoporosis.

Alcohol Abuse

Excessive alcohol use (more than two drinks per day) can affect bone mass. In fact, low bone mass is found in twenty-five to fifty percent of men who seek medical help for their addiction to alcohol.[5]

Alcohol affects hormones such as parathyroid hormone (PTH), calcitonin, cortisol, and growth hormone which are involved in bone growth and remodeling. Heavy or chronic drinking also causes low levels of activated vitamin D which decreases the absorption of calcium from the GI tract. Finally, alcohol abuse can lower levels of testosterone, leading to hypogonadism as discussed above. It's worth mentioning that excessive alcohol intake can also predispose people to falls.

Smoking

Like alcohol abuse, smoking has been shown to be a cause of osteoporosis in both men and women. Nicotine and other

4. "Osteoporosis in Men." National Institute of Arthritis and Musculoskeletal and Skin Diseases. June 2015. Accessed February 01, 2017. https://www.niams.nih.gov/Health_Info/Bone/Osteoporosis/men.asp.
5. Ibid.

toxic substances in cigarette smoke reduce levels of PTH, testosterone, and estrogen and increase levels of cortisol. Smoking reduces the level of vitamin D in the body and decreases the absorption of calcium. Cigarette smoke increases free radicals (destructive cells that cause damage and disease) and oxidative stress in the body and reduces the blood supply to bones (a double whammy!). It also reduces the number of bone-forming cells (osteoblasts) so that less bone is made. Finally, people who smoke tend to weigh less, be less physically active, and to have poor nutrition (all of which contribute to osteoporosis).

Use of Glucocorticoids

The most common cause of secondary osteoporosis in men is the use of glucocorticoid medications. These drugs are also known as "steroids" and are used for the treatment of inflammatory disorders such as asthma, chronic obstructive pulmonary disease (COPD), and rheumatoid arthritis. Glucocorticoid drugs are not the same as anabolic steroids which are male hormones used by athletes to increase muscle mass. Common medications in this class are prednisone, cortisone, and methylprednisolone. These drugs have a direct effect on bone absorption and breakdown, and may reduce absorption of calcium and levels of testosterone.

Bone mass decreases quickly with ongoing use of glucocorticoids. Men taking these drugs for more than three months should talk to their doctor about having a bone mineral density (BMD) test and have testosterone levels checked regularly.

Gastrointestinal Disorders and Poor Nutrition

Several nutrients are essential for bone growth and maintenance including calcium, magnesium, protein (amino acids), phosphorus, and vitamins D and K. People suffering from disorders of the gastrointestinal tract such as celiac disease and inflammatory bowel disease (Crohn's Disease and Ulcerative Colitis) as well as those who have had bariatric (weight loss) or GI surgery may have impaired nutrient absorption. In addition, those with poor diet or anorexia may be at risk for osteoporosis.

Hypercalciuria

Hypercalciuria is a disorder characterized by increased excretion of calcium in the urine. Known causes of hypercalciuria include high levels of parathyroid hormone, sarcoidosis, cancer, Cushing's syndrome, hyperthyroidism, Paget's Disease, and a type of kidney disease called renal tubular acidosis. Sometimes the cause of increased calcium excretion is not known. When excessive amounts of calcium are lost in the urine, bone health may be affected.

Immobilization

Weight bearing exercise is important for maintenance of healthy bones. When people are not able to move due to surgery, disorders like Parkinson's Disease or multiple sclerosis, stroke, or spinal cord injury (to name a few reasons); bone density declines rapidly. It is crucial for people to resume weight bearing exercise like walking as soon as possible after a period of immobility.

The Skeleton in the Closet—Screening and Diagnosis

A physical exam and careful medical history are important for detecting of osteoporosis. Your physician may also want to obtain blood and urine tests. Recommended guidelines for osteoporosis screening in men vary by the organization. The National Osteoporosis Foundation (NOF) and the Endocrine Society recommend testing bone mineral density (BMD) in all men who are seventy years of age or older. In addition, men between the ages of fifty and sixty-nine should be tested if they have risk factors for fractures or have suffered a fracture. The test that is considered the "Gold Standard" for determining bone density is called the dual energy x-ray absorptiometry test (DEXA). The DEXA test measures bone mineral density at the hip and lower spine using a "T-Score." A T-score of -2.5 or less is indicative of osteoporosis. A score greater than -1 is normal. A T-score between -1 and -2.5 shows that bones are becoming weak. Bone density tests are quick, painless, and involve very low amounts of radiation. In fact, you are exposed to more radiation on a cross country flight. DEXA testing should be repeated every one to two years.

Don't Step on a Crack—Osteoporosis and Fracture Prevention

Wouldn't it be better to prevent osteoporosis than to be subject to a painful fracture? If you are a man at risk for osteoporosis, there are some simple steps that you can take to prevent the disease from occurring.

Diet—It's Not Only About Calcium

The saying "we are what we eat" comes into play when we talk about prevention of osteoporosis. A balanced diet with plenty of protein and calories is important when it comes to overall health as well as for osteoporosis prevention. Of course, optimal amounts of calcium, vitamin D, and magnesium are also essential for maintenance of bone density. The Food and Nutrition Board (FNB) at the Institute of Medicine of the National Academies recommends that adult men consume a total of 1000mg per day of calcium up until the age of seventy. Men over seventy should have a total calcium intake of 1200mg per day. Dairy products such as milk, cheese, and yogurt are good sources of calcium as well as leafy green vegetables such as broccoli and kale and bony fish like salmon. Many foods are fortified with calcium. See Table 3 for calcium content of selected food products. Those who don't obtain enough calcium in their diet should take calcium supplements to ensure

Table 3: Calcium Content of Selected Food Items		
Food	**Amount**	**Calcium Content**
Milk	1 cup	300mg
Cottage Cheese	½ cup	65mg
Ice Cream	½ cup	100mg
Sour Cream	1 cup	250mg
Soy Milk	1 cup	300-400mg
Yogurt	1 cup	450mg
Hard Cheese	1 oz	200mg
Parmesan Cheese	1 tbsp	70mg
Swiss Cheese	1 oz	270mg
Broccoli (cooked)	1 cup	180mg
Kale (raw)	1 cup	55mg
Spinach (cooked)	1 cup	240mg
Figs (dried)	1 cup	300mg
Orange juice (fortified)	8 oz	300mg
Almonds	1 oz	80mg
Canned salmon	3 oz	180mg
Canned sardines	4 oz	350mg

adequate intake. Vitamin D is also important for bone health and prevention of osteoporosis. Dietary vitamin D should be 600 International Units per day up to age seventy and 800 IU daily for those over the age of seventy. Good sources of vitamin D include milk, egg yolks, and fatty fish. Men with low levels of vitamin D should receive supplements. Magnesium is also important for bone mineralization as well as other critical functions in the body. It is found in legumes, nuts, grains, and fish. Consult your health care professional before taking any nutritional supplements for osteoporosis prevention.

Exercise

Prevention of osteoporosis offers us another incentive to leave the couch. Weight bearing exercises such as running, walking, and dancing stimulate osteoblasts to strengthen our bones. Swimming is an excellent aerobic exercise, but since it isn't weight bearing, it won't help to prevent osteoporosis. People who have not been active in the past or who are older might want to begin slowly-perhaps with walking. Regular exercise is recommended—thirty minutes of physical activity at least five times weekly is ideal. Weight training is especially important because it not only assists in strengthening our bones, it also increases muscle tone and helps with coordination and balance, which may help to prevent falls.

Falls

Falls can happen at any age and can significantly increase the risk of fractures. It's really important to reduce our risk of falls by taking following steps:

- Wear comfortable shoes that fit properly with non-skid soles; make sure laces are tied

- Avoid walking on slippery surfaces such as snow, ice, and wet floors
- Remove throw rugs or make sure they don't slide by using double-sided tape or slip-resistant pads
- Place electrical cords out of walkways
- Walk in areas that are well-lit inside and outside the home
- Keep a nightlight between your bed and the bathroom
- Use handrails and grab bars in stairwells and bathrooms
- Make sure to have vision checked regularly

Lifestyle Choices

As mentioned in previous sections, it is important to quit smoking and avoid drinking more than two drinks daily for osteoporosis prevention. Excess drinking may also cause falls.

Give That Dog a Bone—Treatment of Osteoporosis

The decision to administer pharmacologic treatment for osteoporosis is based on clinical evaluation, risk of fractures, and bone marrow density measurements. In 2008, the World Health Organization (WHO) developed a tool to assess risk of fractures. This tool is called the fracture risk assessment tool (FRAX). It can be accessed at http://www.shef.ac.uk/FRAX. The fracture risk assessment tool estimates the ten-year risk of major osteopathic fracture and hip fracture by factoring in risk factors like age, weight, sex, history of fracture, smoking, alcohol use, and bone marrow density.

Guidelines recommend treatment for men who are fifty years of age or older, have a T score between -1.0 and -2.5 in the hip, spine, or femoral neck, and have a ten-year risk of twenty percent or higher of developing any fracture or three percent or greater of developing a hip fracture.

Men who are above the age of fifty who are taking glucocorticoids should also be considered for treatment.

Medications for Osteoporosis in Men

Several types of medications are FDA approved for treatment of osteoporosis in men. These include the bisphosphonates, parathyroid hormones, a monoclonal antibody, and testosterone. (Table 4)

Table 4 Medications for Osteoporosis in Men	
Bisphosphonates	Teriparatide (Forteo)
Alendronate (Fosamax)	Monoclonal Antibody
Risedronate (Actonel, Atelvia)	Denosumab (Prolia)
Zoledronic Acid (Reclast)	Testosterone
Parathyroid Hormone	

The Bisphosphonates

These medications include Alendronate (Fosamax), Risedronate (Actonel or Atelvia), and the injectable Zoledronic acid (Reclast). This type of medication works by inhibiting the osteoclasts that break down and remove bone. They are first line for treatment of osteoporosis in men and have been studied the most.

You should take Fosamax and Actonel first thing in the morning on an empty stomach with a full (eight ounce) glass of water. You should take Atelvia just after breakfast with four ounces of water. After taking these medications, wait at least thirty minutes before eating or taking other medications or vitamins, calcium, or antacids. Don't lie down or recline for at least thirty minutes after taking these bisphosphonates. Reclast is given by your health care professional into your vein as an IV medication. The infusion will last at least fifteen minutes. It is usually given once a year for treatment of osteoporosis. Patients may eat before receiving Reclast and should drink plenty of fluids throughout the day.

Most people who take alendronate and risedronate don't experience any serious side effects. Common side effects of alendronate and risedronate include nausea, abdominal pain, indigestion, heartburn (dyspepsia), joint and/or muscle pain, and diarrhea or constipation. Rare side effects of the bisphosphonates include inflammation or ulcers of the esophagus and osteonecrosis of the jaw. Osteonecrosis of the jaw occurs when the jaw bone doesn't receive an adequate blood supply and thereby is also starved of nutrients. It's important for those taking bisphosphonates to have good oral health—brush and floss teeth daily and see a dentist on a regular basis. Some people who take Reclast experience flu-like symptoms within one to three days after receiving the first dose. The symptoms may include fever and joint/muscle pain. Taking acetaminophen usually lessens discomfort.

Parathyroid Hormone

Teriparatide (Forteo) is a synthetic (or man-made) version of parathyroid hormone (which is naturally found in the body and regulates bone growth). Forteo is the only medication that actually stimulates the formation of bone. It is given as a daily injection under the skin for two years. Forteo is a treatment for men with severe osteoporosis or for those whose osteoporosis has not improved with bisphosphonates. Side effects associated with Forteo are mild and include pain at the injection site, dizziness, nausea, and temporary high calcium levels four to six hours after the dose. The use of Forteo is limited to two years due to the possible risk of bone cancer if large doses are given. Patients should rotate injection sites in the thigh or abdominal areas and sit when injecting the medication.

Denosumab (Prolia)

Denosumab (Prolia) is a type of drug known as a monoclonal antibody. Monoclonal antibodies are proteins that bind to a type of cell in the body. Denosumab (Prolia) binds to the osteoclast cells in the body to prevent these cells from breaking down bone. Prolia is a treatment for men with osteoporosis who are at high risk for fractures including those who have suffered from a previous fracture. It is also used for patients who have failed or cannot take other treatments for osteoporosis. It is given as an injection under the skin every six months. Prolia is generally well tolerated by patients, but some side effects can occur such as skin infections and rash. Like Forteo, Prolia can cause a mild, temporary elevation in calcium levels.

Testosterone

According to the latest Endocrine Society guidelines, treatment with testosterone should be given to men with low levels of the male hormone and high risk of fracture. The society also recommends that these men also receive a bisphosphonate or Teriparatide along with testosterone therapy. Studies have shown that testosterone increases bone marrow density in men with low levels testosterone. So far, studies have not assessed the effects of combination therapy of testosterone with bisphos-phonates or other drugs for osteoporosis.

So Whatever Happened to Jeff?

Jeff's physician prescribed calcium supplements along with a once weekly Fosamax tablet. Jeff has stopped smoking and limits his drinking. He recovered from his hip fracture and his bone density is much improved.

The Bottom Line

Osteoporosis is starting to be recognized as a significant health problem for men, but men are still less likely to be diagnosed and treated than females. Men can take steps to prevent bone loss before it happens by eating well, taking calcium and vitamin D supplements, avoiding unhealthy habits like excessive alcohol consumption and smoking, and exercising on a regular basis. Fall prevention strategies can also be an effective way to avoid fractures. Make sure to see your physician on a regular basis and to ask about your risk for osteoporosis. Those at risk can take steps to avoid further loss of bone mass and fractures.

CHAPTER 10

When There's a Pain in the Pelvis That Won't Go Away—When It Really Hurts Down There

contributed by Mindi S. Miller, PharmD, BCPS

Vladimir (aka Vlad), age forty-three, has pain and discomfort in the area under his scrotum. He goes to the restroom more than twenty times a day. He describes the toilet as his new best friend! He gets up five to seven times a night just to urinate. He is worn out and tired upon awakening in the morning and believes that his symptoms are affecting his work and his relationship with his wife. His urine examination is negative, there is no evidence of blood in his urine, nor evidence of a urinary tract infection. He has pain and discomfort with urination. He was treated with multiple medications including antibiotics and muscle relaxants without relief. Vlad, and his wife, are not happy campers.

Chronic pelvic pain (CPP) affects both women and men. Though CPP has traditionally been considered a "women's" disease, men also get pelvic pain. For both sexes, many of the challenges are similar, even with regard to sexually related pain.

But men with CPP face challenges of their own. The diagnosis is often missed because CPP is less common in men than in women, and its symptoms overlap with those of more common conditions in men. CPP has multiple other names including interstitial cystitis, chronic non-bacterial prostatitis, bladder pain syndrome, and pelvic floor dysfunction. For the remainder of this chapter, all diagnoses in men with chronic pelvic pain will be referred to as CPP.

The symptoms for CPP include pelvic pain or a dull ache in the pelvic area, urinary urgency, painful urination, pain with intercourse, and urinary frequency. CPP can affect a man's social life, sleep, and even his ability to function effectively in the work place. Some men have to urinate forty to sixty times a day. It is no wonder that a CPP affected man's best friend is the nearest toilet. Men with CPP often state that they are prisoners of their pelvis. It is not uncommon for men with CPP to awaken multiple times a night because they have a full bladder and pelvic pain.

But these symptoms overlap with other conditions that are more common in men, especially chronic bacterial prostatitis, urinary tract infections, and benign enlargement of the prostate gland (See Chapter 2).

There is a trend today to consider chronic pelvic pain and IC as the same condition. Perhaps the hallmark of CPP is the presence of urinary symptoms, pain and discomfort, with a normal urine examination, a negative urine culture, i.e., no bacteria identified, and no response to antibiotics.

If a man with CPP also has chronic lower urinary tract

symptoms, such as urgency, frequency, nocturia, pain with bladder filling, suprapubic pressure, or painful urination, and does not respond to standard therapies for prostatitis, he may have CPP. Clinical experience suggests that if these patients are treated specifically for CPP, they tend to do better than if they are treated only with typical chronic prostatitis therapies, such as alpha-blockers and/or antibiotics.

It is often difficult to differentiate prostatitis from CPP. With prostatitis, the pain is experienced in the area between the testicles and anus. This area is referred to as the perineum. Prostatitis is characteristically accompanied by urinary symptoms, and these men have an abnormal digital rectal exam with a tender prostate, and also have an abnormal urine examination—if the cause is due to an infection, the urine culture will confirm the infection.

Men are more reluctant than women to say they have pain and may be less apt to share that they have bladder and pelvic pain. You know that macho thing! Sociological studies show that this tendency is ingrained in a man early in life. By age five or six, boys are less likely than girls to express hurt or distress. Little boys are told "don't cry," "be a man," and "suck it up!" Research also shows that men's coping skills are not as well developed as women's. These tendencies mean that men may not seek pain control as soon as they need to, and may need more help developing coping skills in order to be more comfortable sharing their medical concerns with a health care professional.

Causes of CPP

The bladder is shaped like a balloon and is located in the pelvis. It has the ability to expand and contract just like a balloon. It has two tubes, called ureters, which transport urine that is made in the kidneys to the bladder where it is stored. There is another tube, the urethra, located in the penis, which allows the urine to leave the bladder into the toilet when the man has the desire to urinate. Normally the bladder is able to expand and hold fluid or urine without any discomfort. However, in patients with CPP, the expansion stretches the muscles, which irritates the nerves that supply the bladder and causes severe pain. As a result, men with CPP have intense pain as the bladder expands, which also makes them feel that they always have to go to the bathroom. Most men find that the pain in the pelvis will subside after they urinate. In severe cases, even after urinating, the pain will persist.

The best explanation we have of the cause of CPP is that there is a defect in the lining of the bladder that enables a toxic chemical in the urine to penetrate and irritate the muscles and the nerves that supply the bladder, resulting in the urge to urinate. As the lining becomes leaky, scar tissue forms in the muscles of the bladder, which decreases the capacity of the bladder to expand when fluid is added to the bladder from the kidneys. This situation then causes men to have a smaller bladder capacity and to have to urinate more frequently.

No one knows what causes IC/PBS, but doctors believe that it is a real physical problem and not a result, symptom, or sign of an emotional problem. Because the symptoms of CPP are varied, most researchers believe that it represents a spectrum of disorders rather than one single disease.

One area of research on the cause of CPP has focused on the layer that coats the lining of the bladder called the glycocalyx, made up primarily of substances called mucins and glycosaminoglycans (GAGs). This layer normally protects the bladder wall from any toxic contents in urine. Researchers have found that this protective layer of the bladder is "leaky" in about 70 percent of CPP patients and have hypothesized that this may allow substances in urine to pass into the bladder wall where they might trigger CPP directly, or may make these patients susceptible to other chemicals in the urine, including those from foods or beverages.

Along with altered permeability of the bladder wall, researchers are also examining the possibility that CPP results from decreased levels of protective substances in the bladder wall. Reduced levels of GAGs (discussed previously) or other protective proteins might also be responsible for the damage to the bladder wall seen in CPP.

No matter what the mechanism for disruption of the bladder lining, potassium is one substance that may be involved in damage to the bladder wall. Potassium is present in high concentrations in urine and is normally not toxic to the bladder lining. However, if the tissues lining the inside of the bladder (urothelium) are disrupted or are abnormally leaky, potassium could then penetrate the lining tissue and enter the muscle layers of the bladder where it can cause damage and promote inflammation.

Researchers have isolated a substance known as anti-proliferative factor (APF) that appears to block the normal growth of cells that make up the lining of the bladder. APF has been identified almost exclusively in the urine of people suffering with CPP. Research is under way to clarify the potential role of APF in the development of CPP.

Increased activation of sensory nerves (neurologic hypersensitivity) in the bladder wall is also thought to contribute to the symptoms of CPP. In addition, cells known as mast cells within the bladder wall, which play a role in the body's inflammatory response to injury, release chemicals that are believed to contribute to the symptoms of CPP.

Other theories about the cause of CPP are that it is a form of autoimmune disorder (in which the body's own immune system attacks the body) or that infection with an unidentified organism may be producing the damage to the bladder and the accompanying symptoms.

Diagnosis of CPP

It is very important for male patients to have a thorough diagnostic workup. This includes a careful history and a physical exam including a digital rectal exam to check the prostate gland. Urine tests are ordered and include a urinalysis and a urine culture to be certain that there is no blood in the urine and no urinary tract infections. Another test which is commonly performed is a cystoscopy (a look into the bladder with a lighted tube). In approximately five percent of men with CPP, there are characteristic ulcers, referred to as Hunner's ulcers, located in the inner lining of the bladder with a cystoscopy.

Another diagnostic test is hydrodistention of the bladder, or filling the bladder with sterile water to determine the bladder capacity. Hydrodistention is performed under general or regional anesthesia. This workup will help rule out other medical conditions that cause pelvic pain and will help to confirm chronic pelvic pain as a diagnosis.

Urinalysis and Urine Culture

These tests can detect and identify the most common bacteria in the urine that may be causing IC/PBS-like symptoms. A urine sample is obtained either by catheterization with a small tube inserted into the penis, or more commonly by the "clean catch" method. For a clean catch, the man washes his genital area before collecting a sample of urine "midstream" in a sterile container. White and red blood cells and bacteria in the urine suggest an infection of the urinary tract that can be treated with antibiotics. If urine is sterile for weeks or months while symptoms persist, a doctor may consider a diagnosis of CPP.

Culture of Prostatic Secretions

Using a digital rectal examination, the doctor can obtain a sample of prostatic fluid. This fluid is examined under the microscope for signs of an infection such as red and white blood cells, and also can be cultured for bacteria. Prostatic infections can be treated with antibiotics.

Potassium Sensitivity Test

A test known as the intravesical potassium sensitivity test (PST) has been developed to evaluate the leakiness of the protective lining of the bladder. Some experts recommend its use in the evaluation of CPP, but we feel it is antiquated, and the pain and the discomfort are a significant deterrent for most doctors trying to diagnose CPP. At the present time, we are not recommending the use of the PST.

Lidocaine Instillation

Filling the bladder with a solution containing the local anesthetic drug, lidocaine, has been described as an "anesthetic bladder challenge." Improvement of symptoms after lidocaine has been instilled into the bladder suggests CPP. However, this test is not specific for CPP and is not routinely performed.

Cystoscopy Under Anesthesia with Bladder Distension

During cystoscopy, the doctor uses a cystoscope—an instrument made of a hollow tube about the diameter of a drinking straw with several lenses and a source of light—to look inside the bladder and urethra. The doctor will also distend or stretch the bladder to its capacity by filling it with a liquid or inert gas such as carbon dioxide. Because bladder distension is painful in CPP patients, the patient is often given either regional or general anesthesia before the doctor inserts the cystoscope through the urethra into the bladder. Cystoscopy with distension of the bladder with sterile water can detect inflammation (visually or with biopsies), a thick and stiff bladder wall, and Hunner's ulcers. After the fluid has been drained from the bladder, small red spots, called glomerulations, that represent enlarged blood vessels and pinpoint areas of bleeding can be seen in the bladder's lining.

The cystoscopy also allows measurement of patient's bladder capacity—the maximum amount of liquid or gas the bladder can hold under anesthesia (without anesthesia, capacity is limited by either pain or a severe urge to urinate). Most people without CPP have normal or large maximum bladder capacities under anesthesia. A small bladder capacity, due to scarring of the bladder wall, helps to support the diagnosis of CPP.

Cystoscopy is recommended only to exclude other possible causes of symptoms and not as the definitive diagnostic test for CPP.

It is important to note that the distension often performed with cystoscopy may lead to relief of symptoms in some patients with CPP, which generally lasts from several weeks to months following the procedure.

Bladder Biopsy

A biopsy is a microscopic examination of a small sample of tissue. Samples of the bladder and urethra may be removed during cystoscopy and examined under a microscope at a later date. A biopsy helps to exclude bladder cancer, and also may confirm the presence of mast cells or inflammation of the bladder wall that is consistent with a diagnosis of IC/PBS. Nevertheless, there is nothing on the biopsy that can make an absolute diagnosis of IC/PBS. Unfortunately, there are no lab tests to distinguish CPP from non-bacterial prostatitis or urethritis.

We can summarize the diagnosis of CPP by stating that the condition is difficult to diagnose and is often considered a diagnosis of exclusion, and though treatments can reduce symptoms, there's no cure. In most clinical situations, diagnosing the problem with the patients' history, the clinical presentation, and a physical exam is all there is.

Treatment for the Pain "Down There"

Many methods have been tried to provide relief from symptoms of CPP. However, no standard or universally accepted treatment has been found that reliably and invariably resolves this debilitating condition.

Removing the Offending Foods and Fluids from Your Diet

Many men with CPP find that eliminating or reducing their intake of potential bladder irritants may help to relieve their discomfort. The most irritating foods can be summarized as the "four C's." The four C's include carbonated beverages, caffeine in all forms (including chocolate), citrus products, and food containing high concentrations of vitamin C and potassium.

Table 1
Common bladder irritants to avoid
Tomatoes (includes ketchup, salsa, pizza sauce)
Caffeine
Chilies/Spicy foods
Chocolate
Citrus fruits and drinks: Oranges, limes, and lemons
Alcoholic beverages: beer, wine, liquor
Carbonated beverages: soft drinks, soda water, energy drinks
Spicy foods: salsa
Sweeteners: artificial and natural sweeteners
Processed foods
Onions
Cranberries
Vinegar
Raw onions

If you find that these things irritate your bladder, you may also wish to avoid related foods such as tomatoes, pickled foods, alcohol, and spices. Artificial sweeteners may also aggravate symptoms in some men. If you think certain foods make you feel worse, try eliminating them from your diet. Reintroduce them one at a time to determine which, if any, affect your signs and symptoms. Additionally, making your urine less acidic by taking alkalinizing agents such as citrate may improve symptoms of frequency and pain associated with CPP. Citrates, which include potassium or sodium citrate, tricitrates, and citric acid, either alone or in combination (Bicitra, Citrolith, Oracit, Polycitra, Urocit-K), are usually used to prevent certain types of kidney stones. But because they make the urine less acidic, they may help relieve bladder pain in men with CPP. These are available by prescription.

Most men can identify the culprits by conducting a simple elimination diet. We suggest that you stop eating all of the foods on the bladder irritant list shown in Table 1. We recommend that you adhere to this elimination diet for two weeks. If the bladder symptoms improve, then you can be sure that one or several of the foods or liquids was irritating your bladder. Then begin by eating just one food from the bladder irritant list food list. If you don't develop any symptoms of pelvic pain or increased urgency or urinary frequency within 24 hours, then you can be reasonably certain that this is not one of the foods or fluids that is effecting your CPP. You can then add one new food or liquid each day until you have identified the offending foods that are guilty of making you so miserable.

Oral medications for treating CPP

Because the cause of CPP is unknown, numerous empirical therapies have been tried. Oral medications have given relief to some men and are commonly the first-line treatment. The only FDA approved drug for the treatment of interstitial cystitis/painful bladder syndrome (IC/PBS) is polysulfate sodium or Elmiron. Elmiron is a very weak blood thinner or anticoagulant. Elmiron probably works by coating the inner lining of the bladder that becomes damaged by the disease, and helps to restore the inner lining so that the bladder lining no longer leaks. Perhaps Elmiron prevents toxins in the urine from penetrating the inner lining and irritating the nerves under the bladder lining that cause the severe pain, and urinary frequency and urgency. Unfortunately, Elmiron may take between three to six months to improve urinary symptoms and pelvic pain. A decrease in pain is often one of the first reported positive effects, with improvement in frequency and urgency of urinary symptoms occurring after pain relief. It appears that about 35–40 percent of CPP men who take Elmiron notice marked improvement. Part of the reason that Elmiron may take so long to work is that it often has to work to undo months or years of damage to the bladder's inner lining, resulting from CPP. Elmiron itself does not heal the bladder wall; instead, it offers a layer of protection to the bladder while it heals, which is often a drawn-out process. As such it's best to stay on Elmiron for at least a six-month trial period, as it may take that long to see any positive effects.

Elmiron is a very safe drug and only a few men with CPP experience side effects such as diarrhea or nausea. A very rare side effect is hair loss, which occurs in about three percent of the men who use Elmiron. Fortunately, the hair loss is temporary and the hair will grow back once the medication is discontinued.

Other oral medications include anticholinergic drugs such as Ditropan, Detrol, Vesicare, Toviaz, Enablex, Mybetriq, and Sanctura, which are used to decrease bladder spasms. There are also bladder analgesics such as pyridium, which turns the urine red, or Prosed, which turns the urine blue. These medications allow you to pee in technicolor! Both of these drugs can temporarily relieve the pain or burning associated with urination. Over-the-counter forms of phenazopyridine hydrochloride (Azo-Standard, Prodium, and Uristat) may provide some relief from urinary pain, urgency, frequency, and burning. Higher doses of the phenazopyridine (Pyridium) are available but require a doctor's prescription.

Other medications that may reduce the symptoms of CPP include sedatives for those men who have trouble falling and staying asleep. Anti-depressants have also been effective. Some of the most commonly used sleep aids include tricyclics such as amitriptyline (Elavil), doxepin (Sinequan) and imipramine (Tofranil), given at bedtime. These agents may be started at very low dosages and gradually increased until symptom relief is obtained or the side effects become bothersome. Use of hydroxyzine (Atarax), an antihistamine, is also effective especially if the biopsy of the bladder shows an excess of mast cells, which release histamine, responsible for the pain in the bladder.

Other Oral Medications Used to Treat CPP

Other oral medications that may be used to treat CPP include low doses of antidepressants of the tricyclic group such as amitryptiline (Elavil). It is believed that tricyclic antidepressants can help reduce the hyper-activation of nerves within the bladder wall. The anti-seizure medication gabapentin (Neurontin, Gabarone, Gralise, Horizant, Fanatrex FusePag) also has been used to treat nerve-related pain and has sometimes been used to treat the pain of CPP. Oral antihistamines, such as hydroxyzine, also may be prescribed to help reduce allergic symptoms that may be worsening the patient's CPP symptoms.

Leukotrienes, which are substances produced by some immune system cells and mast cells, promote inflammation. Drugs that block leukotrienes are new and are being used in the treatment of asthma and allergy. They include the prescription medicines:

- Montelukast (Singulair)
- Zafirlukast (Accolate)
- Zileuton (Zyflo)

New evidence has implicated leukotrienes in inflammation of the bladder in men with CPP. The leukotriene receptors have been found in the bladder muscle in patients with CPP. Patients who took Singulair for three months showed significant reductions in urinary frequency and pelvic pain.

Medication Placed into the Bladder to Sooth the Pain "Down There"

If you take medication by mouth, the drug must be absorbed through the gastrointestinal tract, then introduced into the blood stream, and finally filtered by the kidney and excreted in the urine, where the drug can do its job against the abnormal lining of the bladder. However, you can perform a blood stream-kidney bypass by inserting the medication directly into the bladder. By placing a tiny catheter or tube through the urethra, the bladder is drained of urine and then medication is placed in the bladder where it heals the altered lining of the bladder. The medication only remains in the bladder for a few minutes and the man goes to the restroom and urinates the remaining medication into the toilet. This treatment option has almost no side effects because only a very small amount of the medication is absorbed from the bladder. Medication instilled into the bladder is usually not painful or even uncomfortable.

Many agents have been instilled into the bladder in attempts to relieve symptoms. Today, the most commonly used medication that is instilled in the bladder is dimethyl sulfoxide (DMSO) or RIMSO. DMSO is approved by the FDA for the treatment of CPP. Although DMSO is an anti-inflammatory agent, it seems to have some pain relieving properties as well. As is true of other therapies for CPP, there is no way to predict which men may or may not respond to DMSO.

DMSO treatment regimens vary and have included instillation weekly for six weeks, and men especially respond to a "booster" dose every month. There is no uniformity of choice regarding treatment schedules. Most likely, each doctor uses a slightly different approach; as yet, there is no "right" way to use DMSO. It is probably safe to say that if a man with CPP does not experience relief within a few treatments, further instillation of DMSO is unlikely to be effective. The most common side effect of DMSO is a complaint of a garlic-like taste and odor after instillation. Most men will find that sucking on a peppermint candy after the treatment alleviates the garlic taste.

Following the initial course of treatment, some men achieve long-term remission, but most men will eventually relapse. Additional treatment schedules for those who relapse vary but usually consist of instillation every four-to-six weeks. There are some men who learn how to insert the small catheter and instill the DMSO in the privacy of their homes and can use the treatment whenever symptoms occur without having to make a doctor's appointment. Sometimes DMSO is combined with other medications such as heparin, steroids, bicarbonate, and a local anesthetic (xylocaine) as a bladder treatment. A bladder "cocktail" consisting of heparin, lidocaine, which is a local anesthetic, and sodium bicarbonate has been effective in some men. This cocktail is given several times a week for several weeks and many CPP sufferers respond immediately to this treatment.

Botox® Can do More Than Take the Wrinkles out of Your Face— it's Good for Pain "Down There"

Botox®, which has been used for years to wipe away those wrinkles around the mouth and forehead, has being investigated for injection into the bladder muscle to reduce bladder spasms in men with CPP who have not responded to oral medication or bladder instillations. The procedure is usually done in the operating room where a very tiny needle is inserted through a cystoscope and the bladder muscle is injected at multiple locations with Botox®. Improvement is seen within a few days and if there is improvement, Botox® injections will often last from four to six months and then the procedure can be repeated.

Botox® is made from the botulism toxin. It has been modified and has been utilized in FDA-approved procedures such as for the treatment of wrinkles. By injecting Botox® into his bladder, her nerves were "put to sleep," calming their harmful influence on the bladder. Unfortunately, Botox® does not cure the problem and must be repeated every three to twelve months.

Physical Therapy

Although physical therapy (PT) is well known for helping men recover from orthopedic surgery, PT also helps men with soft tissue injuries. PT has been shown to be helpful to ease muscular spasm in the pelvic floor. Although it might be difficult to think about going through this therapy because pelvic floor muscle massage requires a digital rectal exam, PT can offer tremendous relief. PT also includes learning how to avoid activities that exacerbate your symptoms by releasing muscle trigger points and using techniques that keep the pelvic muscles relaxed, and even learning to treat yourself so you can do the exercises in the comfort and privacy of your own home. It helps to involve your spouse or significant other in PT, as he or she can help massage muscles and trigger points you cannot reach yourself. We suggest you find a pelvic floor physiotherapist who has experience in treating men with CPP. Pelvic floor physiotherapists sometimes combine the PT technique with biofeedback. (We have provided a list of websites at the end of this chapter to locate a physical therapist near you who treats CPP.)

You Must Prevail When All Else Fails

The vast majority of men with CPP can be helped with medications, drugs placed in the bladder, and hydrodistension. However, there are few unfortunate men where nothing seems to help. They have tried all of the conservative approaches and have seen multiple physicians and experts who treat CPP, and yet the men remain miserable and incapacitated by the disease. In these few men, a surgical option may be the only solution. Attempts have been made to sever the nerves to the bladder that appear to be associated with the pain and discomfort. Occasionally, some men will have a part or even the whole bladder removed. For those men with CPP who have a very small bladder that holds only a small volume of urine, they may have their bladder capacity enlarged by attaching a piece of intestine to the bladder in order to increase the its capacity.

Nerve Stimulation

If you have tried diet changes, exercise, and medicines and nothing seems to help, you may wish to think about nerve stimulation. This treatment sends mild electrical pulses to the nerves that control the bladder.

At first, you may try a system that sends the pulses through electrodes placed on your skin. If this therapy works for you, you may consider having a device put in your body. The device delivers small pulses of electricity to the nerves around the bladder.

For some patients, nerve stimulation relieves bladder pain as well as urinary frequency and urgency. For others, the treatment relieves frequency and urgency, but not pain. For still other patients, it does not work.

The treatment consists of electrical pulses, which block the pain signals carried in the nerves. If your brain doesn't receive the nerve signal, you don't feel the pain.

Mild electrical pulses can be used to stimulate the nerves to the bladder—either through the skin or with an implanted device. The method of delivering impulses through the skin is called transcutaneous electrical nerve stimulation (TENS). With TENS, mild electric pulses enter the body for minutes to hours, two or more times a day either through wires placed on the lower back or just above the pubic area—between the navel and the pubic hair. The electrical pulses may increase blood flow to the bladder, strengthen pelvic muscles that help control the bladder, or trigger the release of substances that block pain.

TENS is relatively inexpensive and allows people with CPP to take an active part in treatment. Within some guidelines, the patient decides when, how long, and at what intensity TENS will be used. It has been most helpful in relieving pain and decreasing frequency of urination. Smokers do not respond as well as nonsmokers. If TENS is going to help, improvement is usually apparent in three to four months.

A person may consider having a device, Inter-Stim, or sacral neuromodulation, implanted that delivers regular impulses to the bladder. A wire is placed next to the tailbone and attached to a permanent stimulator under the skin. The FDA has approved this device to treat CPP when other treatments have not worked.

Acupuncture

Another option being tested as an alternative treatment modality is the use of acupuncture. This consists of two to three treatments each week for several months. The mechanism of action on the use of acupuncture for the treatment of CPP is not known. Perhaps the symptoms of frequency and urgency are alleviated because pressure in the bladder and urethra is reduced by an action of the acupuncture needles on blocking excessive nerve impulses. However, if the acupuncture treatment can adjust the nerve impulses to the bladder, it may be possible to alleviate some of the sense of urgency that occurs with small amounts of urine in the bladder, and it may also reduce some of the inflammatory processes that occur in the bladder lining.

Coping Tips for Patients with CPP

There is probably no problem in the pelvis that requires a more sympathetic family and significant other. Emotional support is very important in coping with ongoing pain and the necessity of going to the restroom so frequently. Family and friends can often supply the support. However, you are not alone. There are excellent support groups that meet on a regular basis, and you can also find support groups online that are usually very helpful. (See resources at the end of the chapter for support groups near you.)

Here are Some Tips to Remember:

- Find a health care team that is sympathetic and helpful.
- Understand that your health care team does not know all the answers and may be as frustrated as you are. Often, a team of health care professionals is needed to address problems encountered by

someone suffering from CPP. Psychosocial support is often necessary as depression also can be present.

- Stay in touch with family and friends. Don't become isolated.
- Involve your family in treatment decisions.
- Remember that CPP is only one part of your life. Don't allow it to become all of your life.
- Talk to others about their experiences and ways of coping.
- Timed voiding is important—voiding by the clock rather than going to bathroom whenever you feel the urge. Over time, the bladder is reeducated as the intervals between voids are lengthened.
- Wear loose clothing. Avoid belts or clothes that put pressure on your abdomen.
- Avoid activities that you know will flare your CPP: certain body positions, very bumpy rides, etc.
- Reduce stress. Try methods such as visualization, meditation, biofeedback, and low-impact exercise.
- Stop smoking as the active ingredient in tobacco, nicotine, may worsen any painful condition, and smoking is harmful to the bladder.
- Try pelvic floor physiotherapy. Gently stretching and strengthening the pelvic floor muscles, possibly with help from a pelvic floor physiotherapist, may reduce muscle spasms. Pelvic floor physiotherapists sometimes combine this technique with biofeedback.
- Treatment of the muscles and connective tissue in the pelvis by a highly trained physical therapist is highly beneficial for patients with CPP and other

pelvic pain conditions. Instead of focusing on squeezing these muscles, such as during Kegel exercise, therapists teach the patient how to relax these muscles. During therapy, the specialist focuses on release of muscle tension and pain by manipulating the connective tissue and muscle in certain directions. The treatments are often supplemented with mild electrical, heat, and/or cold treatments.

The Rest of Vlad's Story

So what happened to Vlad, who was held hostage by his chronic pelvic pain? Vlad had seen more than nine health care providers including three urologists, two primary care doctors, a naturopath, a chronic pain management specialist, a neurologist, a psychiatrist, and a dietician. He was suffering on a daily basis describing his pain as a seven to eight out of ten. He often had to take opiate medication for pain relief. He was experiencing depression and loss of productivity at his work. Finally, Vlad was able to find a compassionate physician who specializes in chronic pelvic pain and reassured him that his problem was not a psychiatric one and that he wasn't crazy. Vlad was able to work with the pain management specialist to discontinue the use of opiate pain medication. Vlad was advised to stop bladder/prostate irritants such as caffeine, carbonated beverages, citrus fruits, and juices. Vlad joined one of the support groups that is mentioned at the end of this chapter so he knew he wasn't alone, and that there were others who had a similar problem. Vlad had several bladder instillations of DMSO by his urologist and was able to use meditation and self-talk to relieve some of his urinary symptoms. Over a period of several months Vlad was able to gain control of his symptoms and described his pain as two to three, which made life worth living.

The Bottom Line

Chronic pelvic pain is a common problem in many primary care practices, and very common in urologic practices. Still, it is probably under diagnosed. Because the symptoms are non-specific and can be very confusing, men are sometimes thought to have psychogenic problems or are thought to have a bacterial infection, and are treated with multiple courses of antibiotics despite the absence of evidence of bacterial infection. The key to the correct diagnosis is awareness of the condition and its characteristics. The bottom line is that achieving a complete cure for chronic pelvic pain may not be possible; however, it is in reach of most men to achieve long-term remission, or certainly a reduction in symptoms that makes life worth living.

CHAPTER 11

Incontinence: Diapers—You Don't Have to Depend on Depends™

Ebenezer, a sixty-nine-year-old man, had prostate cancer and had his prostate removed through a robotic procedure (see Chapter 3). After surgery, he had a problem of urinary incontinence and had to wear diapers. The incontinence was a source of despair and even mild depression. It affected nearly every aspect of his life. He and his wife were unable to sleep in the same bed because of the incontinence and sexual intimacy was not even remotely possible.

From the time we are toddlers until we become older and infirm, we tend to take control of urination for granted. Seldom do we have embarrassing accidents between ages three and eighty-three. However, as men age, the muscles holding the urine in place until we can reach a restroom are weakened, and loss of urine becomes one of aging's most troublesome and devastating experiences. This chapter will discuss the problems that many middle aged and older men have, and what can be done about it.

Loss of urine is no laughing matter. It affects nearly fourteen million American adults. Let's put the myth aside that incontinence only affects women. Incontinence is surprisingly common in men of all age groups. Losing urine is only half as frequent in men as in women, which represents quite a sizeable number of sufferers. Unfortunately, men with incontinence rarely discuss their problem with their physician, so the necessary attention is not paid to finding a cure to the condition, or even to managing the problem.

Surprisingly, twenty-five percent of men aged forty or below reported incontinence at least once during the previous twelve months. At least thirty percent of men over age forty, and thirty-six percent of men age sixty to seventy had at least one incident of incontinence in the past year.

A Brief Review of the Male Plumbing System

In order to understand the problem of incontinence, a brief review of the male plumbing system is in order. The kidneys function to filter out liquid waste from the blood. The kidneys then combine this liquid waste with water and transport the urinary fluid to the bladder, where the urine is stored until it's time to pee. While the bladder muscle squeezes, the external sphincter relaxes, allowing the urine to be released and passed through the urethra, the tube in the penis that transports the urine to the outside of the body. (Figure 1)

If there is any problem with the bladder muscle not being able to contract, if it contracts too frequently without the

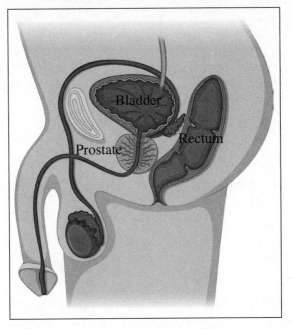

Figure 1: Male Pelvic Anatomy *(Shutterstock)*

owner's permission, or if there is a weakness of the sphincter that does not maintain the urine inside the bladder until it is time to empty the bladder of urine, then the problem of urinary incontinence will occur.

There are multiple causes of urinary incontinence. It can be as simple as not drinking enough water. It could be as serious as an inflamed bladder wall from a urinary tract infection. Several diseases can bring about incontinence, such as multiple sclerosis, diabetes, stroke, and Alzheimer's disease. Prostate problems can hamper urination. Men who had had prostate surgery, like Ebenezer, or bladder surgery, or who were taking medications for urinary problems, had an increased likelihood of being incontinent (two to three times more likely). Table 1 is list of drugs that may cause urinary incontinence.

Table 1
Drugs that can cause urinary incontinence.
Alpha-adrenergic blockers
Antihistamines
Angiotensin-converting enzyme inhibitors
Calcium channel blockers
Diuretics
Analgesics/opioids (Percocet, Oxycontin)
Skeletal muscle relaxants
Antidepressants
Anticholinergics
Sedatives

Drinking coffee or tea or taking prescribed medications with adverse side effects can aggravate your bladder. Certain drugs (e.g., diuretics, alcohol) have no direct action on the urinary tract, but may contribute to incontinence by increasing urine production or impairing nervous system function.

Not surprisingly, daily, frequent incontinence is associated with deterioration in a man's quality of life. For instance, emotional health, social relationships, physical activity, and travel are all less satisfactory for incontinent men. The problem even impacts a man's relationship with his significant other, as well as his family and friends.

Unfortunately, only a third of the men with incontinence discuss the problem with their physicians. However, three-quarters of them express an interest in having a full evaluation and treatment of the problem, if it were offered.

It can be concluded that male incontinence is a real problem across all age groups, and that it affects men's quality of life. Unfortunately, the sufferers do not often discuss it with their physicians. There is clearly much room for improvement in its diagnosis and management.

Types of Incontinence

There are five common types of urinary incontinence: stress incontinence, overflow incontinence, urge incontinence, mixed incontinence, and functional incontinence.

Stress Incontinence

Stress incontinence in men is usually a result of prostate surgery for either benign prostate enlargement, or surgical removal of the prostate gland for cancer of the prostate. If you have stress incontinence, you leak small amounts of urine when you cough, sneeze, exercise, or put pressure on your bladder, and the muscle or sphincter is unable to hold the urine in the body.

Urge Incontinence

Urge incontinence occurs when your bladder suddenly contracts and expels urine. You get an urge to urinate even though you know you emptied your bladder not long before. You urinate, and then get the urge again a half-hour later. Urge incontinence often comes in waves. It may not bother you all morning, for example, but it becomes insistent mid-afternoon. In the course of a few hours you may feel the urge four or five times.

Mixed Incontinence

Mixed incontinence occurs when you have symptoms of two or more distinct types of incontinence, usually stress and urge incontinence. The symptoms of mixed incontinence are loss of urine with coughing and sneezing (stress component), and having to go to the restroom immediately with the first desire or urge to urinate (urge component). It is important for your doctor to know the type of incontinence as the treatments for each kind of incontinence may differ.

Overflow Incontinence

Overflow incontinence is a form of urinary incontinence, characterized by the involuntary loss of urine from an overfull urinary bladder, often in the absence of any urge to urinate. Overflow incontinence occurs when you are unable to completely empty your bladder; this leads to overflow, which leaks out unexpectedly. You may or may not sense that your bladder is full. The leakage, which can cause embarrassment and discomfort, is not the only problem. Urine left in the bladder is a breeding ground for bacteria. This can lead to repeated urinary tract infection and possibly kidney damage. The most common cause in men is an enlarged prostate, which impedes the flow of urine out of the bladder and results in constant dribbling of small amounts of urine both during the day and night.

Functional Incontinence

Functional incontinence is a form of urinary incontinence in which a man is usually aware of the need to urinate, but for one or more physical or mental reasons he is unable to reach the rest room in time and incontinence, or loss of urine, occurs. Often, the cause of functional incontinence is a problem that keeps the man from moving quickly enough to get to the bathroom, pull down his pants to use the toilet, or transfer from a wheelchair to a toilet. The causes include problems with mobility such as arthritis, back pain, or neurological problems, such as that which occur with Parkinson's disease, or after a stroke.

The Evaluation

A voiding diary helps the doctor achieve an accurate diagnosis and helps create a treatment plan. The doctor can achieve a better understanding of a man's problem of incontinence if the man has kept a voiding diary for just a few days prior to his examination by the doctor. Many men provide an unclear voiding history; therefore, a voiding diary can be helpful. (Figure 2)

Voiding Diary

Date	Time	Urinated in Toilet	Small Accident	Large Accident	Fluid Intake	Circumstances of Accident

Figure 2: Voiding Diary

The simplest voiding diaries ask patients to record the frequency of incontinence episodes, but diaries also can be used to assess the situations in which incontinence occurs, which can help clarify the type of incontinence. For example, the diary may reveal leakage during times of increased abdominal pressure, suggestive of stress incontinence, or dribbling that is indicative of overflow incontinence. Men with stress incontinence usually wake once or not at all at night to urinate. However, patients with urge incontinence usually wake more than twice and as often as every hour.

The Physical Examination

The physical examination can identify clues that help lead to an effective diagnosis. The doctor will examine the abdomen and see if the bladder can be felt above the pubic bone, which is suggestive of overflow incontinence or a bladder that is not completely empty. The lower legs are examined for joint impairment, which might indicate incontinence secondary to an inability to reach the toilet in a timely fashion (functional incontinence). If there is swelling of the lower extremities, this may be an indication of heart disease, and that the incontinence is a result of excessive fluid that overwhelms the kidneys and bladder. All men require a prostate examination or digital rectal examination.

In addition to the history and the voiding diary, the physical exam requires minimal testing, which includes urinalysis, a urine culture, a cystoscopy, and perhaps urodynamic testing. The latter test includes measuring the bladder capacity, the flow of urine from the bladder to the outside of the body, and any urine that is retained in the bladder. A few blood tests are done to check kidney function, calcium, and glucose levels.

Treatment of Urinary Incontinence

There are some simple life-style changes that every man can do that may diminish his urinary incontinence.

Low impact sports such as cycling, yoga, or elliptical machine exercises are ideal activities for keeping fit without affecting a sensitive bladder condition.

Abdominal workouts such as sit ups, crunches, or plank kicks place a lot of pressure on the pelvic floor. Opt for alternative exercises where breathing or the position itself supports the pelvic floor.

Avoid heavy lifting. Lifting heavy objects is particularly bad for the pelvic floor and back. Ask for help instead.

Pelvic Floor Exercises

Pelvic floor exercises and targeted Pilates and yoga exercises that strengthen the pelvic muscles can be particularly helpful. By practicing at least three times a day, they can help strengthen the pelvic floor muscles and give more control when needed.

Kegel exercises may help strengthen the muscles in your pelvic floor. This allows you to delay urinating until you reach a toilet.

You may have thought that Kegel exercises were something only women do. In fact, the muscles that are strengthened with Kegel exercises are the same in both sexes. (For more information on Kegel exercises, see resources at the end of the chapter.)

Biofeedback is often prescribed for men with stress incontinence following prostate gland surgery for prostate cancer. Men will often have a weakness in the pelvic floor muscles after surgery that requires treatment and strengthening. Therapists will often utilize bladder training, biofeedback, pelvic floor muscle exercises, and electrical muscle stimulation—which is not painful, but encourages contraction and relaxation of the muscles that are responsible for holding the urine in the bladder.

Men typically attend in-office appointments two to three times per week for a duration of twelve weeks. Treatment requires a digital rectal exam evaluating the external and internal tissues of the pelvis, especially those surrounding the urethra. This type of therapy is effective in some men, especially if they complete the course of treatment and do the exercises between treatment sessions.

Drink Just Enough

There's no need to avoid drinking in order to reduce the urge to visit the bathroom. Limiting water intake makes urine more concentrated, which boosts the chances of bladder irritation.

Just say no to caffeine. Caffeine, alcohol and carbonated drinks could be your new worst enemies. Try limiting coffee, tea and carbonated beverages for a week or two as they can irritate a sensitive bladder.

Set your alarm on your smart phone to alert you to use the restroom. Your bladder is trainable. If you need to pass water frequently and need to rush to the restroom, ask your doctor about a daily schedule for building up the bladder's holding capacity. Remember, allow your bladder to empty completely each time you go to the toilet.

A growing number of pads for day and night use as well as absorbent underwear and bed pads are available at pharmacy chains, on the Internet, and at discount stores. Wearing a diaper or pull-ups may be the difference between being confined to your home and feeling able to leave the house and participate in social activities.

Condom Catheter

A condom catheter is a treatment option for men who have constant loss of urine or are unable to receive any warning that urination is imminent. A condom catheter consists of a condom with the tip of the condom cut off, so it could be attached to a drainage tube and collecting bag which is often worn on the leg or thigh.

Condom catheters are made in three sizes: small, medium and large, and usually contain adhesives or skin sealants, which are non-irritating to the underlying skin. The skin sealant protects your skin from perspiration and urine irritation; when you remove the condom catheter, the layer of skin sealant is removed, not the external layer of your skin.

Penile Clamp

A penile clamp uses a hinged, rigid frame that supports two pads and a locking mechanism. It controls leakage by applying constant pressure upon the penis and compresses the urethra, the tube in the penis that transports urine from the bladder to the toilet.

The penile clamp cannot be used for long periods of time and should not be placed on the same location of the penis since the clamp decreases the blood supply to the underlying skin and urethra and can result in damage to the urethra. We recommend that the penile clamp be moved up or down the shaft of the penis every two hours and must be removed while sleeping. For the most part, the penile clamp is a temporary solution and a man with incontinence should not become dependent on this treatment option.

Medical Management

Medication is a possibility for men with incontinence related to prostate gland enlargement, or for men with urge incontinence.

Incontinence secondary to an enlarged prostate can be managed with medications to relax the prostate muscles, or alpha-blockers and medications used to reduce the size of the prostate gland, finasteride, and dutasteride (see Chapter 2). There are other options for the enlarged prostate gland that don't respond to medications, which are described in Chapter 2.

For men who have urge incontinence, there are medications to relax the bladder and thus reduce the urges and the incontinence. These are called anticholinergic medicines such as oxybutynin, which control bladder muscles and increase bladder capacity. Antidepressant medicines, such as imipramine, may also be effective and help with bladder control.

There are very few medications that are effective for stress incontinence. If a man has both stress and urge incontinence, then anticholinergics may be helpful with the urge component of the incontinence. Also, the antidepressant medication, imipramine, may be helpful with mild stress incontinence.

Surgical Options

Male Slings

A small sling made of soft synthetic mesh can be implanted inside the body to provide support to surrounding muscles. This can help to keep the urethra closed, especially when coughing, sneezing, and lifting. (Figure 3)

Artificial Sphincter

The artificial urinary sphincter (AUS) is placed inside the body. A saline-filled cuff compresses the urethra and prevents urine from leaving the bladder. The AUS also contains a pump in the scrotum, which can be activated to remove fluid from

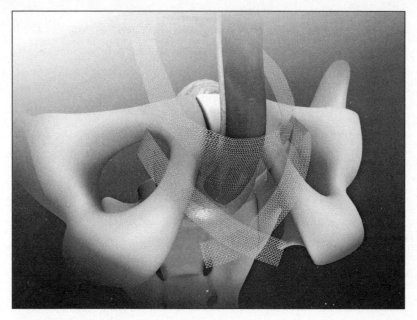

Figure 3: Male Sling *(Coloplast)*

the cuff, and releases the compression on the urethra to allow urination when the man feels his bladder is full. The AUS is designed for use after prostate surgery.

So What Happened to our Miserable Ebenezer?

He used diapers and an external condom catheter for several months, practicing Kegel exercises diligently every day. He gained considerable control over his incontinence. He was able to avoid urinary accidents by making an effort to urinate before leaving his home. He used a tiny pad for any dribbling that occurred when his bladder was full. He returned to his bedroom with his wife and his depression lifted. He was taking oral medication for his erectile dysfunction and his PSA tests after surgery were 0.01ng\ml, which means he was cured of his *prostate cancer. Let's just say that Ebenezer got his life back and both he and his wife were happy campers.*

Bottom Line

Most men's issues with urinary incontinence can be improved with simple lifestyle changes and pelvic floor exercises as well as by finding the right products that help make a man socially comfortable.

Most men with incontinence can be helped and many can be cured with exercises, medication, and surgery. Sufferers should not suffer in silence, but speak to their physicians and health providers so that more attention is paid to their problem. Just remember, you don't have to depend on Depends!™

CHAPTER 12

Sexually Transmitted Infections— An Ouch in The Crouch

When Charlie, age twenty, woke up in his dorm room, he had a very strong urge to urinate—more than most mornings. Expecting relief as soon as he reached the bathroom, he was surprised by the painful burning as he emptied his bladder. When he looked down, he noticed a yellowish discharge in the front of his underwear. Since it was the weekend, the campus health center was closed. He did not know where to turn. He certainly could not call his parents. Although he was sexually active, he did not have a serious girlfriend in whom he could confide. His best thought was to turn to the Internet for a solution.

Introduction

When most people think of sexually transmitted diseases, they think of Herpes and HIV. As if that was not enough to worry about, there are a host of viruses, bacteria, and other organisms that can be transmitted sexually. Not only can these infections be spread through vaginal and anal contact, but also through oral and manual (hand) contact. In some cases, these infections can cause some very bothersome symptoms. At other times, symptoms can be delayed. Often, untreated sexually transmitted infections (STIs) can cause irreversible damage to the genitourinary and other organ systems. The good news is that most of these diseases are preventable. It all starts with knowledge.

Prevention by Intention

Other than abstinence, the best way to prevent acquiring a sexually transmitted infection is to commit to a monogamous relationship with an uninfected partner as determined by their physician. When a monogamous relationship is not possible, sex with as few partners as possible is the next best thing.

Of course the most common method of prevention (when the status of infection is unknown) is condom use. However, condoms are only effective when used properly. First and foremost, any penile contact should involve a condom. This hard-and-fast rule not only applies to vaginal and anal contact, but also oral and manual contact. Both semen and vaginal secretions of infected individuals can be highly contagious. Even "casual" contact—either before or after sexual intimacy—can lead to unsuspected exposure to these secretions.

Although somewhat controversial in the public's eye, the Centers for Disease Control has well-established recommendations for vaccination.[1] Every man should be vaccinated for

---

1. "Immunization Schedules." Centers for Disease Control and Prevention. March 06, 2017. Accessed August 01, 2017. https://www.cdc .gov/vaccines/schedules/hcp/child-adolescent.html.

HPV and Hepatitis B early in life. These vaccinations will be discussed later in this chapter.

Early Detection

Medical attention should be sought at the first sign of symptoms. These symptoms can include pain (genital or pelvic), skin lesions, urethral discharge, vaginal discharge, pain with intercourse (especially for females), and pain with urination. Any of these concerns should be shared with your sexual partner or partners so that they can be properly treated. In some circumstances, diagnosis can be easier in your partner, thereby avoiding a missed diagnosis.

With the exception of a mutually monogamous relationship in which both partners are confirmed to be uninfected, regular testing is the hallmark of prevention and early detection. Most people do not realize that many sexually transmitted infections, such as HIV and syphilis, may not cause any symptoms, particularly at the onset. More importantly, some of these infections can cause irreversible damage to the genitourinary tract and other organ systems. Appropriate testing can help prevent some of these long-term consequences.

After the Fire—Late Complications of Sexually Transmitted Infections

Several of the sexually transmitted viruses are now known to cause a variety of cancers. Human papillomavirus (HPV) can cause cervical, mouth, throat, and anal cancer. Hepatitis B and C can cause liver cancer. The Centers for Disease Control recommends vaccination for both HPV and Hepatitis B. Untreated human immunodeficiency virus (HIV/AIDS) can lead to an assortment of cancers and death. Early treatment of HIV can help prevent cancers associated with this disease.[2]

Since dealing with an STI is a couples issue, a man should also be concerned about the well-being of his partner—especially when it can affect future fertility. Sexually transmitted infections have a much higher risk of future fertility problems in women than in men. These infections can cause pelvic inflammatory disease, also known as "PID." Chlamydia and gonorrhea are the most common causes of PID and can lead to scarring and blockage of the fallopian tubes (the tubes that carry the eggs to the uterus). Similarly, a man's tubes can also develop scar tissue from untreated gonorrhea and put him at risk for future fertility problems. Furthermore, any active sexually transmitted infection during pregnancy can threaten the well-being and life of the fetus.

The Hidden Culprits

All sexually transmitted infections can be treated, and most can be cured. Knowledge is power when it comes to fighting these bacteria, viruses, and other organisms. Unfortunately, you may not notice any symptoms with many of these infections. For this reason, you should see a physician for testing when starting a sexual relationship with a new partner or if there are any concerns of having a sexually transmitted disease. What follows is a guide to the most common sexually transmitted

---

2. "STIs and Cancer-." Home -. Accessed August 01, 2017. http://www .ashasexualhealth.org/stis-and-cancer/.

infections, including typical symptoms, methods of diagnosis, and the standard treatments.

Genital Warts

Human papilloma virus, or HPV, is the causative organism of genital warts. These warts can appear anywhere along the penis, scrotum, anus, inner thigh, vagina, and cervix. This virus has dozens of subtypes, some of which do not cause the classic wart associated with this disease. The virus can be transmitted with oral, genital, anal, and even skin-to-skin contact. Other than visual inspection, the diagnosis can be made with biopsy of the skin lesion or swabbing of the cervix.

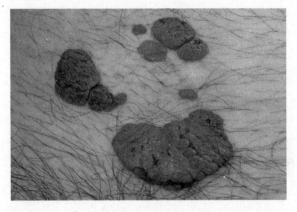

Figure 1: Genital Warts *(Wikimedia Commons)*

Not all types of HPV are harmful, but some can lead to penile, anal, mouth, and throat cancer. Prevention, early detection, and proper treatment are paramount in avoiding these cancers. The CDC recommends that young men ages eleven to twenty-one be vaccinated for HPV.[3]

3. Centers for Disease Control and Prevention. Sexually Transmitted Disease Surveillance 2015. Atlanta: U.S. Department of Health and Human Services; 2016.

Many of the subtypes of HPV are not symptomatic and are eradicated by the body's immune system. Visible skin lesions usually appear within two to three months following infection and are most commonly treated with chemical or laser destruction, depending on the extent and location. In some cases, your physician may prescribe one of two available topical creams, Condolox® (Podofilox®) or Aldara® (imiquimod). These creams can take many weeks and can cause irritation of the surrounding skin. Although the creams clear the warts in roughly 50 percent of patients, 20–50 percent will recur. Laser destruction is usually performed in an operating room setting.

The Clap and the Clam

Although caused by different organisms, gonorrhea and chlamydia often co-exist in an infected individual. In fact, even when only one of these diseases is detected, treatment is given for both. The most common symptoms are burning in the urethra (worse with urination) and a milky or yellowish discharge from the urethra. Like most sexually transmitted infections, these organisms can be transmitted through vaginal, anal, and oral sex. Over three hundred-fifty thousand cases of gonorrhea and over 1.5 million cases of chlamydia are reported each year.[4]

Urethral swabbing and testing of any urethral discharge usually confirms the diagnosis. It is very important that your partner also be treated and that you refrain from sexual contact until both you and your partner have both been fully treated. Most men with gonorrhea have symptoms consisting of burning

---

4.   "2015 Sexually Transmitted Diseases Treatment Guidelines." Centers for Disease Control and Prevention. January 25, 2017. Accessed August 01, 2017. https://www.cdc.gov/std/tg2015/default.htm.

on urination, frequency of urination, and a urethral discharge. Taking into account the growing resistance of these organisms to certain antibiotics, the CDC recommends dual therapy with a single intramuscular dose of ceftriaxone (250 mg) plus a single oral dose of azithromycin (1 gm). Once again, both organisms, gonorrhea and chlamydia, must be treated, even when only one is detected by testing.[5]

The Gift That Keeps on Giving

Herpes viruses come in many forms and can cause a variety of diseases including chicken pox, shingles, cold sores, mononucleosis ("mono"), and genital herpes. Unlike HPV, herpes is not usually associated with a future risk of cancer (although the subtype that causes mono is associated with lymphoma).

The herpes viruses that are transmitted sexually are called herpes simplex and are classified in two types. Herpes virus type 1 (HSV-1) is usually associated with oral blisters or "cold sores," whereas herpes virus type 2 (HSV-2) is usually associated with genital lesions. However, either type can occur in or around the mouth, penis, anus, and rectum. The lesions are characterized by painful blisters. Although these lesions are treatable with antiviral medications, there is no cure to prevent recurrences. For those with recurrent disease, a characteristic pain or sensation can precede blister formation by several days. During both this "prodrome" phase and when the lesions are visible is the time during which you are most contagious. Herpes is very contagious and can be transferred by any affected skin

5. "Genital HSV Infections." Centers for Disease Control and Prevention. June 08, 2015. Accessed August 01, 2017. https://www.cdc.gov/std/tg2015/herpes.htm.

not covered by a condom. Although there is no sure-fire way to prevent transmission to an uninfected partner, any person with HSV should consult their physician for advice on how to minimize this risk.

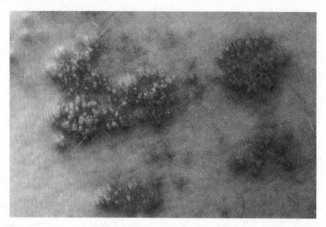

Figure 2: Herpes Skin Lesion *(Wikimedia Commons)*

The diagnosis of HSV can usually be made by visual inspection of the lesion by your physician. At times, your physician may swab the sore for laboratory testing. You should also observe any changes in the appearance of the blisters and any sensations or discomfort that preceded the lesion. Often, the blister starts with a small red patch that contains multiple pinhead-size blisters that soon coalesce to form one blister. These are usually painful and tender to the touch, but some cause no pain (especially inside the vagina or rectum). The blisters will then break open and form an ulcer before forming a scab during the healing phase.

As mentioned, herpes simplex is treatable but not curable. In fact, once contracted, it lies dormant in your nervous system for your lifetime. In some cases, years pass between exposure and your first outbreak. Treatment focuses on quicker healing of sores, lessening the severity and frequency of outbreaks, and minimizing transmission of the virus to an uninfected partner (again, not fool-proof). Three oral medications are currently available—acyclovir, valacyclovir, and famciclovir. Although topical antiviral creams have been used in the past, they have since been shown to be ineffective. The CDC recommends a should read "a seven to ten day course of" oral antiviral medications in all patients with their initial outbreak. Shorter courses can be used for subsequent outbreaks. In fact, for those who can identify the "prodrome" of abnormal sensations that precede blister formation, a one to three day regimen can be very effective if started promptly. In these cases, a supply of pills should be on hand for effective "self-start" therapy.[6]

Another approach is suppressive or preventative (prophylactic) therapy. Daily low-dose administration (reduced by as much as half) may be appropriate for those with frequent outbreaks. The frequency of outbreaks often decreases with time, and therefore the need for daily dosing should be re-evaluated on an annual basis with your doctor. However, the risk of transmission to an uninfected partner can be dramatically reduced with continuous daily dosing, particularly in conjunction with condom use. These benefits should be carefully weighed against the disadvantages of long-term use such as cost, side effects, and future drug resistance.

The Deadliest Catch—Human Immunodeficiency Virus (HIV)

Over the past few decades, we have come a long way in the management of HIV (human immunodeficiency virus),

---

6. "What's New in the Guidelines? Adult and Adolescent ARV Guidelines." National Institutes of Health. July 14, 2016. Accessed August 01, 2017. https://aidsinfo.nih.gov/guidelines/html/1/adult-and-adolescent -treatment-guidelines/0/.

the virus that causes AIDS. However, we have still not found a cure or an effective vaccine. This virus infects cells in our blood stream that are responsible for a major part of the function of our immune system. As a result, those individuals infected with HIV are more susceptible to other infections and some cancers, many of which are otherwise rare. Although many people have managed their HIV infection for ten or more years with medications, this disease should be considered potentially fatal.

Infection with HIV often occurs alongside other sexually transmitted infections. In fact, anyone infected with HIV should be tested for all of the other diseases mentioned in this chapter (among others). In addition, the presence of another STI will make you more susceptible to HIV infection since they break down our natural barriers.

Regular testing should be done for those with multiple sexual partners, a new partner, and other high-risk exposure (needles, tattoos, health care providers). Considering the gravity of such an infection, trust in a relationship should not be underestimated.

Although there is no cure for this virus, tremendous advances in treatment have been made over the last few decades. The CDC currently recommends a "cocktail" of three different antiretroviral drugs from at least two different drug classes. Many infected individuals can go many years with undetectable viral counts. Despite these low counts, care must be taken to avoid exposure to others.[7]

---

7. "Syphilis Statistics." Centers for Disease Control and Prevention. December 19, 2016. Accessed August 01, 2017. https://www.cdc.gov /std/syphilis/stats.htm.

Syphilis-The Great Imitator

No discussion of sexually transmitted infections is complete without the inclusion of syphilis. It was called the Great Imitator because it was often confused with other conditions. Often, not much attention is given to syphilis because it is not as common as other STIs and because it is curable. However, we are starting to see a rise in the incidence of this disease.[8] In fact, the CDC recommends that all pregnant woman be tested since syphilis can be easily passed from mother to baby.[9] More importantly, if not treated in the early stages, syphilis can lead to some serious health and life-threatening conditions.

Syphilis is divided into four stages—primary, secondary, latent, and tertiary. Primary syphilis is characterized by small, round, and painless bumps (called "chancres") in the area where the organism (the "spirochete" bacterium *Treponema pallidum*) originally entered the skin. Transmission usually occurs via the penis, vagina, anus, rectum, or mouth. Whether or not treatment is received at this time, the chancres last three to six weeks. During or shortly after the primary stage is healing, secondary syphilis will manifest with a flat, itchy rash that can occur on any part of the body (most commonly on the palms or on the soles of the feet). Often, this rash can have minimal symptoms and can be difficult to diagnosis unless syphilis is being considered. Like primary syphilis, secondary syphilis will resolve whether or not it has been treated. The latent stage of syphilis can last for years without any signs or symptoms. However, if it

8. "Syphilis During Pregnancy." Centers for Disease Control and Prevention. July 27, 2016. Accessed August 01, 2017. https://www.cdc.gov/std/tg2015/syphilis-pregnancy.htm.

9. Chosidow, Olivier. "Scabies and pediculosis." *The Lancet* 355, no. 9206 (2000): 819–26. doi:10.1016/s0140-6736(99)09458-1.

was never properly treated, it can unpredictably lead to the dangerous tertiary stage. Tertiary syphilis can have life-threatening effects on the brain (neuro-syphilis), eyes (ocular syphilis), and other internal organs.

As already mentioned, the key to diagnosis is having the suspicion of the possibility of syphilis. Often the patient's description of events and risk factors can be helpful, and at times the spirochete can be sampled with a swab to confirm its presence with laboratory testing. However, the diagnosis is usually made with a special blood test called an RPR. Syphilis is simply treated with penicillin. For those allergic to penicillin, doxycycline is usually effective. Of course, all sexual partners from the time of initial infection must also be treated.

The Crabs

There is nothing like an unrelenting jock-itch to make you "crabby." However, unlike the traditional jock-itch that is caused by a fungus (as is athletes' foot), crabs is caused by lice, a parasitic insect. It is very contagious—not only through sexual contact, but also through non-sexual contact with clothes and other inanimate objects, such as pillows. Condoms are usually not effective since the pubic hair is exposed (even with "man-scapes"). Three million cases are reported in the US each year.[10]

Diagnosis can usual be made by identifying the small insects with the naked eye or easily with a magnifying glass. Treatment is simply a one-time application of an over-the-counter lotion, 1% permethrin, to the entire body with special attention to areas with hair (including the eyebrows). Your pharmacist can

---

10. "Viral Hepatitis." Centers for Disease Control and Prevention. August 17, 2017. Accessed September 01, 2017. https://www.cdc.gov /hepatitis/index.htm.

help guide you on selection and use of this product. Also, it is best to re-apply the lotion seven days later to treat any lice that may have hatched following the treatment, but prior to them being mature enough to reproduce. Even though the lice do not survive longer than twenty-four hours away from their human host, it is important to thoroughly clean any clothing, sheets, towels, or surfaces that may contain lice in order to prevent reinfection shortly after treatment (or infection of others).

Trichomoniasis

Trichomoniasis is caused by a parasite and can manifest in itching, burning, malodorous urethral discharge, or no symptoms at all. Treatment is simply a single oral dose of metronidazole 2 gm (Flagyl) for both partners.

Hepatitis

When talking about sexually transmitted infections, most people do not think about liver disease or hepatitis. However, in addition to transmission through blood (needles and transfusions), the various types of hepatitis can be passed via food (especially Hepatitis type A), saliva, and other body fluids. In fact, with most types, sexual contact is one of the more common causes of transmission. Since some of these viruses often co-exist with HIV, a person diagnosed with one of these viruses should be tested for the other.

Common symptoms include fatigue, nausea, fever, and yellowing of the skin and the whites of the eyes. Hepatitis type A (HAV) is usually transmitted through poorly kept or poorly processed food (seafood being the most common) but can also be transmitted sexually. This type usually goes away on its own. For those at higher risk, a vaccination is available. This vaccination is often given to global travelers who go to areas that have a high incidence of Hepatitis A. Incorporation of HAV vaccination into the childhood vaccination schedule is recommended by the CDC.

Hepatitis B (HBV) is transmitted through blood transfusions, contaminated needles, and sexual contact. It is usually treated by careful observation of liver function, monitoring of blood viral levels, and other supportive care. The CDC recommends vaccination of all children, health care workers, and those at higher risk for infection. The CDC hopes that the entire population be vaccinated for both HAV and HBV so that these menacing viruses be eradicated from our communities. Although there is no cure for HBV, there has been some promising results with treating chronically infected patients with immunotherapy to control the disease.

Hepatitis C (HCV) has the same risk factors, transmission methods, and potential consequences as HBV. However, there is a cure for HCV that is FDA-approved and commercially available (direct-acting antivirals).

Hepatitis D (also called Hepatitis Delta or HDV) interestingly must co-exist with Hepatitis B in order to survive. Unfortunately, the combination of both of these viruses together gives a much higher rate of permanent liver damage (70–80 percent) and death (20 percent). Therefore, vaccination for HBV prior to infection with either of these viruses is the most effective way to prevent HDV.

For any of the types of hepatitis, prevention is best achieved with proper vaccination, avoidance of contaminated needles, and sensible sexual practices. If you are exposed to a person or substance at risk for hepatitis, seek medical attention immediately. In some cases, postexposure prophylaxis can stop an infection in its tracks before it has a chance to establish itself in your body.

So Whatever Happened to Charlie?

Unfortunately, his search for an answer on the internet did not yield a home remedy. However, he did find an urgent care center nearby that was open on the weekend. After hearing Charlie's symptoms, the physician suspected that Charlie suffered from urethritis due to a sexually transmitted infection. The doctor placed a small swab into Charlie's urethra in order to take a sample of the secretions for laboratory analysis. The initial examination under the microscope revealed a unique type of white blood cells, polymorphonuclear leukocytes. Since these are characteristic of gonorrhea, he was treated with "dual therapy" of a ceftriaxone injection and a single oral dose of azithromycin.

A few days later, the urgent care center called Charlie to let him know that the culture results revealed gonorrhea and chlamydia, both of which should be covered by the antibiotics that he received. In fact, Charlie's symptoms were almost gone. During the last couple of months, he only shared sexual intimacy with one partner. He kindly shared this information with her so that she could be properly examined and treated.

The Bottom Line

Prevention of sexually transmitted infections begins with an open and honest conversation between two partners. Prior to any sexual activity (vaginal, oral, or anal), both partners should undergo testing. Ideally, both partners should commit to having sex only with one another.

In situations outside of a committed relationship, you should be sexually active with as few partners as possible and always use a condom. Prompt attention to any symptoms, testing between sexual relationships, and appropriate vaccination are the hallmarks to prevention in these circumstances.

CHAPTER 13

Diet and Nutritional Supplements for "Down There" Health

contributed by Mindi S. Miller, PharmD, BCPS

Sam is a forty-seven-year-old truck driver who is five foot eight and weighs 270 pounds. He suffers from high blood pressure that is controlled with medication. Sam is on the road most of the day and admits that he doesn't have time to think about his diet. He eats hamburgers, French fries, and pizza frequently. He also enjoys a few beers nightly once he gets to his destination. Sam's wife told him that he should take caffeine powder as a supplement for weight loss, but he wants some advice on proper nutrition.

You Are What You Eat

As a health care professional, one of the most common questions that I receive is "What should I eat to stay healthy, prevent diseases, and build muscle?" Another common question is "What dietary supplements should I take?" This chapter will examine the foods and supplements that are beneficial for men's health and supplements that should be avoided.

Diet

If you are like most people, you have tried various diets to stay healthy, lose weight, or build lean tissue. Fad diets (like the cabbage soup diet, low fat diets, low carbohydrate diets, and others) may help with weight loss temporarily because they tend to restrict calories. However, as soon as we resume "normal" eating, the weight may come back, or even worse, we may gain more weight. For men (and women) to be healthy, we should eat a balance of carbohydrates, protein, and fat. Let's take a closer look at these macronutrients.

Carbohydrates

Carbohydrates are the body's main source of energy. More than half of our calories should come from "healthy" carbohydrates. However, not all carbohydrates are healthy. For example, sugar is a simple or refined carbohydrate. These types of carbs cause the body to increase insulin levels in the blood, which has been linked to weight gain. Simple or refined carbohydrates contain only one or two sugar molecules and are broken down and released into the blood stream quickly. Other refined carbohydrates include white flour, potatoes, and white rice. A healthy diet includes "complex" carbohydrates which are made up of three or more sugar molecules. These take longer for the body to break down, so the sugar is released into the blood stream more slowly. Complex carbohydrates also provide fiber, which is essential for overall and "down there" health. Examples of complex carbohydrates include whole grain breads, cereals, and pasta, oats, brown rice, fruit, and vegetables.

Protein

Proteins are important nutrients and are found in every cell in the body. The human body contains between thirty thousand and fifty thousand different proteins.[1] Protein is made up of subunits called amino acids. When we digest protein, it is broken down by enzymes (which are themselves proteins) into amino acids and small chains of amino acids called peptides. The amino acids and peptides are utilized by the body to make enzymes, hormones, neurotransmitters, DNA, and antibodies, along with forming muscles, bones, hair, skin, and nails (to name just a few structures). Protein also makes new cells and helps the body to repair damaged ones.

The recommended dietary allowance (RDA) of protein is 0.8 grams per kilogram of body weight.[2] This RDA can be considered the minimum amount of protein an adult needs on a daily basis. If you are active or are stressed by illness, you may need more protein in your diet. Those with kidney disease may require less protein because protein is eliminated by the kidneys. To determine your RDA for protein, multiply your weight in pounds by 0.36. For example, a sedentary man who weighs 180 pounds will need a minimum of sixty-five grams of protein daily. Unfortunately, there is a wide range of opinions on the "ideal"

---

1. Michael T., Joseph E. Pizzorno, and Lara Pizzorno. *The Encyclopedia of Healing Foods.* Place of publication not identified: Time Warner International, 2006.
2. Pendick, Daniel. "How much protein do you need every day?" Harvard Health Blog. June 19, 2015. Accessed January 11, 2017. https://www.hsph.harvard.edu/news/.

amount of protein. The National Institute of Medicine recommends that Americans get 10 to 35 percent of daily calories from protein. Protein is found in meat, seafood, poultry, eggs, beans and peas, nuts and seeds, and dairy and soy products. Like carbohydrates, not all proteins are healthy. Red meat and processed meats (like pepperoni, salami, bologna) are high in saturated fat and sodium and are linked to cardiovascular disease, diabetes, and cancer. Healthier protein sources include seafood, white meat poultry with the skin removed, eggs, low fat dairy products, and vegetarian options such as soy, beans, seeds, and nuts.

Fats

About twenty years ago, low or no-fat diets were popular. Baked goods without fat were flying off the supermarket shelves. However, it didn't take long for most people to realize that they weren't losing weight or becoming healthier on these types of diets. Most of the fats in these products were replaced by sugar.

Fats are actually an essential part of a healthy diet. These nutrients provide energy, assist in absorption of some (fat soluble) vitamins, and maintain healthy hair and skin, among other functions. Like protein and carbohydrates, some fats are healthier than others. Saturated and trans-fats have been shown to raise bad (LDL) cholesterol and lower good (HDL) cholesterol. The American Heart Association recommends limiting saturated fat found in meat and some dairy products to under 6 percent of total calories. Trans-fat is considered to be the most unhealthy type of fat. Trans-fat is manufactured by adding hydrogen to vegetable oil, which causes the oil to become solid at room temperature. Trans-fat is also known as partially hydrogenated oil and allows processed foods such as donuts, cookies, frosting, margarine, chips, and fried foods to have a longer shelf life. Saturated and trans-fats should be replaced with the heart-healthy unsaturated fats. Unsaturated fats include polyunsaturated fatty acids and monounsaturated fats. Healthy fats can lower bad (LDL) cholesterol and raise good (HDL) cholesterol levels. Polyunsaturated fats are found mostly in vegetable oils. Omega-3 fatty acids are a type of polyunsaturated fat that is found in fatty fish such as salmon, mackerel, and trout and also in walnuts and flaxseed. Monounsaturated fats contain vitamin E (an antioxidant) and can also reduce the risk of heart disease. This type of healthy fat is found in olives, nuts, avocados, sesame and pumpkin seeds, and olive, canola, and peanut oil.

Is There a Pill for That?—Dietary Supplements

Over 70 percent of Americans take dietary supplements, and we spend more than $30 billion a year on these products.[3, 4] Over half of these consumers take the products for overall health, and about a third of consumers take the products to fill a perceived nutritional gap in their diet.

A well-balanced diet with the components described above is the best way to obtain all the nutrients necessary to stay healthy. However, some people may get a benefit from using supplements.

3. "Americans Spend $30 Billion a Year Out-of-Pocket on Complementary Health Approaches." National Center for Complementary and Integrative Health. August 02, 2016. Accessed March 15, 2017. https://nccih.nih.gov/research/results/spotlight/americans-spend-billions.

4. "Supplement Use Among Younger Adult Generations Contributes to Boost in Overall Usage in 2016-More than 170 million Americans take dietary supplements." Council for Responsible Nutrition. October 27, 2016. Accessed March 17, 2017. https://www.crnusa.org/newsroom/supplement-use-among-younger-adult-generations-contributes-boost-overall-usage-2016-more.

Dietary or nutritional supplements include such ingredients as vitamins, minerals, herbal products, enzymes, and amino acids.

In this section, we'll discuss what dietary supplements may be beneficial for men for overall health as well as "down there" health. In addition, we'll cover some products that should be avoided. It's important to check with your physician before adding a supplement to your regimen. Some of these products may be harmful to you or may interfere with your other medications.

Supplements for Men's Overall Health

The following supplements are beneficial for men (and women) for general health and wellbeing.

Multivitamins

Experts disagree (as they tend to do in the scientific community) on the value of taking a daily multivitamin. Most studies of people who took multivitamins failed to show a benefit. A relatively large study examined six thousand men over the age of sixty-five who took either a daily multivitamin or a placebo for over ten years. There was no benefit on memory loss for those who took the vitamin. Another study examined 1,700 people who had a previous heart attack. After about five years, those taking a multivitamin did not have a reduced risk of a second heart attack. Two studies showed that there may a small reduction in cancer risk from use of a daily multivitamin. An additional study showed that multivitamins may lower the risk of developing cataracts.[5]

---

5. "Supplement Use Among Younger Adult Generations Contributes to Boost in Overall Usage in 2016-More than 170 million Americans take dietary supplements." Council for Responsible Nutrition. October 27, 2016. Accessed March 17, 2017. https://www.crnusa.org/newsroom/supplement-use-among-younger-adult-generations-contributes-boost-overall-usage-2016-more.

So, what's the bottom line on whether we should be taking a daily multivitamin? There is no evidence that supports use of a multivitamin for everyone, but some men should consider it:

- Those who don't eat an optimal diet for any reason.
- Those who are strict vegetarians who may not be getting enough vitamin B12, iron, calcium, and zinc in their diet.
- Those who have restricted diets for weight loss or other reasons.
- Those who are recovering from surgery or have a serious illness that prevents normal digestion of nutrients.
- Those on medications that impair normal digestion of nutrients. For example, proton pump inhibitors and H2 blockers may prevent B12, vitamin C, calcium, magnesium, and iron from being digested.

Vitamin D

Up to 77 percent of American have a deficiency of vitamin D.[6] Vitamin D is a fat-soluble vitamin and is nicknamed "the sunshine vitamin" because it is produced by the skin in response to direct exposure of the sun's ultraviolet rays. We also ingest some vitamin D in foods such as salmon, sardines, egg yolks, and shrimp. Other foods (such as milk, cereal, and orange juice) may be fortified with vitamin D. Our bodies

---

6. Lefevre, Michael L. "Screening for Vitamin D Deficiency in Adults: U.S. Preventive Services Task Force Recommendation Statement." *Annals of Internal Medicine* 162, no. 2 (2015): 133. Accessed March 20, 2017. doi:10.7326/m14-2450.

synthesize vitamin D and we also eat foods that contain vitamin D, so why do so many Americans have low vitamin D levels? Many of us avoid the sun and wear sunscreen to protect our skin. Correctly applied sunscreen can decrease vitamin D production by more than 90 percent.[7] In addition, most of us don't get enough vitamin D from the foods that we eat.

Vitamin D is an important nutrient for both men and women for several reasons. It regulates the absorption of calcium and phosphorus which are needed for bone growth and strengthening (see chapter on osteoporosis). Vitamin D plays a role in proper functioning of our immune system. In addition, it has been shown to reduce risk of heart disease, multiple sclerosis, diabetes, and influenza. Vitamin D may prevent prostate, colon, and other types of cancers as well.

Because it is difficult to get enough vitamin D naturally, supplements do make sense for many adults. Check with your health care provider to make sure that taking vitamin D is beneficial for you. Current guidelines recommend 600 international units (IU) per day for those under the age of 70 and 800 IU daily for older men. Many experts recommend 800 IU to 1,000 IU for most adults and doses up to 4000 IU daily are considered safe. Vitamin D supplements are available in two forms: vitamin D2 (ergocalciferol) and vitamin D3 (cholecalciferol). The D3 form is considered to be more active and is the preferred supplement.

7. "Vitamin D and Health." The Nutrition Source. May 26, 2015. Accessed March 15, 2017. https://www.hsph.harvard.edu/nutrition source/vitamin-d/.

Calcium

As discussed in the osteoporosis chapter, calcium is vital for bone health and prevention of osteoporosis. Most men should consume a total of one thousand to 1,250 mg of calcium daily depending on age. Dietary sources of calcium are best and include dairy products, leafy green vegetables, bony fish, and fortified food and drinks. Men who don't get enough calcium in their diet should consider taking a calcium supplement if recommended by their health care provider. However, too much calcium may be harmful. Some research has linked excess calcium from supplements to kidney stones, prostate cancer, and heart disease, but more evidence is needed. Calcium citrate may be the preferred type of calcium because it is absorbed by the body better than other types of calcium supplements.

Fish Oil (Omega-3 Fatty Acids)

Fish and fish oil are sources of omega-3 fatty acids, which are important in the formation of every cell in the human body. Omega-3 fats are known as "essential fats" because they can't be produced in the body; we must get these nutrients from food. The two most important omega-3 fatty acids are eicosapentaenoic acid (EPA) and docosahexaenoic acid (DHA). These fats support heart, brain, and nerve cell function and assist in healing of tissues. They also have anti-inflammatory properties.

Because omega-3 fats are essential, fish oil has been promoted as a miracle prevention and cure for many disorders. In fact, this supplement is used daily by almost 20 million Americans—up to ten percent of us.[8] The benefits of fish oil

---

8. "7.8% of U.S. adults (18.8 million) used Fish oil/Omega-3 fatty acids." National Center for Complementary and Integrative Health. August 11, 2016. Accessed March 20, 2017. https://nccih.nih.gov/research/statistics/NHIS/2012/natural-products/omega3.

are controversial, but they lower triglyceride levels and protect against heart disease and stroke, especially in those who have already been diagnosed with these diseases.[9] Omega-3 fats may also be helpful for joint and eye health. Other potential uses such as prevention or treatment of depression, diabetes, cancer, memory loss, and ADHD have not been supported by large scale studies. One study showed a potential correlation between higher levels of omega-3 fatty acids and prostate cancer, but this study has been criticized by many experts.

The best way to ensure adequate intake of omega-3 fatty acids is to eat fish at least twice weekly. Good sources of omega-3s include trout, salmon, sardines, herring, tuna, and mackerel. Men who don't eat the recommended amount of fish and those with known heart disease should ask their health care provider about fish oil supplements. These products do have some potential side effects such as belching, nausea, bad breath, and heartburn. Side effects can be reduced by taking these supplements with food. In addition, large doses of fish oil can increase your risk of bleeding, especially if you have a bleeding disorder, bruise easily, or are taking a medication that thins your blood such as Coumadin or a nonsteroidal anti-inflammatory (NSAID).

Supplements for "Down There"

The following section discusses supplements used for two common disorders in men.

---

9. O'Connor, Anahad. "Fish Oil Claims Not Supported by Research." *The New York Times*. March 30, 2015. Accessed March 20, 2017. https://well.blogs.nytimes.com/2015/03/30/fish-oil-claims-not-supported-by-research/.

Benign Prostatic Hyperplasia (BPH)

Benign prostatic hyperplasia, or an enlarged prostate gland, is common in men as they age. Symptoms of BPH include frequent urination, waking at night to urinate, a sudden urge for urination, weak stream, and dribbling. Several types of medications are FDA approved for treatment of BPH, and many men opt for surgery to relieve symptoms. Dietary supplements are also marketed for treatment of symptoms of BPH. The American Urological Association (AUA) does not support the use of these products because of a lack of evidence of effectiveness and possible lack of safety. The most commonly used supplements are listed below:

- Saw Palmetto
- Pygeum africanum
- Beta-sitosterol
- Rye Grass Pollen

Stop Hanging Around—Erectile Dysfunction (ED)

Erectile Dysfunction is the inability to obtain or maintain an erection. The incidence of erectile dysfunction increases with age, and this disorder negatively affects quality of life for those who suffer from it. "Failure to launch" can happen to anyone from time to time, but chronic ED may be caused by physical problems like diabetes, heart disease, high blood pressure, or low testosterone levels; emotional problems like depression or stress; or substances such as alcohol, medications, illicit drugs, or cigarettes. The most important piece of treating ED is to find out the potential cause. Fortunately, ED may be treated effectively once the cause is known. All of us know about the little blue pill and similar medications, but do dietary supplements work just as well? Natural products have been used in African, Chinese, and other cultures for centuries. However, these products haven't been well-studied for the treatment of ED. Safety may also be a concern. Notify your health care provider before taking any supplements for ED. As with drugs for BPH, the American Urological Association (AUA) does not support the use of these products. Some of these products are discussed below:

Yohimbine

Yohimbine is a chemical found in the bark of the West African yohimbe tree. The few studies of yohimbine in men with ED found small benefits in some men, but most experts believe that it doesn't work any better than a placebo. The most common side effects of this supplement include increased blood pressure, tremors, anxiety, and stomach upset. It also may interact with medications. Yohimbine is on the list of supplements that you should avoid.

L-Arginine

L-Arginine is an amino acid found in meat, fish, and poultry which causes increased blood flow in the body. This supplement is considered to be possibly effective for the treatment of ED. Side effects are uncommon but can include nausea, diarrhea, and worsening of asthma symptoms.

Dehydroepiandrosterone (DHEA)

Dehydroepiandrosterone is a natural hormone that is used by the body to produce testosterone. Like testosterone, DHEA levels decrease as men age. Some evidence shows that DHEA may be effective for ED. This supplement may cause acne but is otherwise safe. Do not use this product if you have prostate cancer.

Ginseng

Ginseng has been used for sexual performance in the Chinese culture for centuries. It works to promote relaxation of smooth muscle in the penis and also increases levels of the neurotransmitter dopamine in the brain. Small studies suggest that ginseng may be effective for ED, but larger trials are needed. This supplement may cause nervousness, insomnia, headaches, dizziness, or upset stomach. It also might interfere with drugs for diabetes and blood thinners.

Horny Goat Weed (Epimedium)

This supplement gets the prize for the best name. It is another herb that has been used by the Chinese for many years to assist with sexual performance. Although studies in rats (not goats) look promising, no studies have been performed in humans.

Supplements to Avoid or "Buyer Beware"

Most of us assume that natural vitamins, minerals, and supplements are beneficial to our health and provide nutrients that we may be missing in our diet. We also assume that products available at our local health food store are safe at the very least. Unfortunately, this may not be the case. The Federal Drug Administration (FDA) reviews all prescription and over the counter drug products for efficacy and safety. However, dietary supplements are regulated differently. Manufacturers of dietary supplements are not required to obtain FDA approval before marketing a nutritional supplement. The manufacturers themselves are responsible for the safety of their product. As a result, these products are considered safe until they are proven to be unsafe. So, what's the bottom line on supplements? Most dietary supplements are safe. However, you should take some steps to make sure that you're getting a good quality product. First, look for a quality seal from one of the four organizations that performs testing and inspections of supplements and the plants where they are manufactured. These include ConsumerLab, Natural Products Association, NSF International, and US Pharmacopeia. Supplements that display a seal from one of these companies contain the ingredients listed on the label, are manufactured properly, and don't include toxic contaminants. You may also want to contact the manufacturer to ask about any research that has been done on the product and reported side effects. Finally, check the FDA website to make sure that the supplement has not been recalled.

Consumer Reports has published a list of supplement ingredients to avoid. These substances are potentially harmful depending on a person's medical history, amount of ingredient contained in the product, and the length of time the substance is taken. The fifteen harmful ingredients are discussed in the following sections:

1. Aconite (Wolfsbane, Monkshood)

This product is used to reduce inflammation and joint pain. It has also been used to treat heart failure. However, aconite contains a fast-acting poison that causes serious side effects such as nausea, vomiting, paralysis, weakness, breathing problems, and possibly death. This product should never be taken by mouth or applied to the skin.

2. Bitter Orange (Methylsynephrine)

Bitter orange is made from the fruit, leaf, and peel of a citrus plant. The supplement is a stimulant that is used as an appetite suppressant for weight loss and for various other disorders including upset stomach, nasal congestion, diabetes, and chronic fatigue. It is available as an oil and can be applied to the skin for fungal infections like athlete's foot and jock itch or inhaled as an aromatic oil. Bitter orange is also found in marmalades and liquors and in cosmetics and soaps. The herb is probably safe in the amounts found in food products and for topical use on the skin. However, the supplement can cause serious side effects like high blood pressure, cardiac arrest, heart rhythm abnormalities, and headaches. Bitter orange is a stimulant and can interact with other drugs that excite the central nervous system.

3. Caffeine Powder (1,3,7 trimethylxanthine)

Caffeine seems like a safe substance since many of us drink coffee, tea, and soda which contain caffeine. This supplement is used to improve mental alertness and attention, to lose weight, and for various medical conditions such as pain, asthma, and migraines. Caffeine is probably safe for most adults when used appropriately. However, this product can be harmful in high doses. It can cause insomnia, nervousness, nausea and vomiting, stomach irritation, anxiety, chest pain, ringing in the ears, seizures, heart arrhythmia, cardiac arrest, and possibly death. Caffeine also interacts with other stimulant drugs.

4. Chaparral (Creosote Bush, Greasewood)

Chaparral is a flowering plant found in the desert. It has been used by Native Americans for infections and diseases such as tuberculosis, sexually transmitted diseases, and snakebites. It has also been used for weight loss, cancer, skin rashes, and even the common cold. However, this supplement has been associated with liver and kidney damage. It should be avoided.

5. Coltsfoot (Coughwort, British Tobacco)

Coltsfoot is also a flowering plant in the daisy family that resembles dandelions. It is found in Europe and Asia. This supplement has been used for respiratory problems such as bronchitis, asthma, cough, and sore throat. This product contains toxic substances called pyrrolizidine alkaloids (PAs) which have been linked to liver damage and possibly cancer. Coltsfoot is unsafe and should be avoided.

6. Comfrey (Blackroot, Slippery Root, Blackwort)

Comfrey is a plant with a black root, large leaves, and small purple or cream colored flowers. It is native to Europe. Comfrey has been used as a tea for all sorts of ailments including ulcers, heavy menstrual periods, diarrhea, cough, chest pain, cancer, and sore throat. It has also been applied to the skin for bruises, joint inflammation, and wounds. Comfrey also contains pyrrolizidine alkaloids (PAs) which can cause liver damage and possibly cancer.

7. Germander (Teucrium chamaedrys)

This plant is found in the Mediterranean. It is used for fever, as a digestive aid, for weight loss, and to freshen breath. The use of germander is unsafe, possibly leading to liver damage and hepatitis.

8. Greater Celandine

Greater celandine is a perennial plant in the poppy family. This herb is used for various digestive problems such as upset stomach, constipation, and irritable bowel syndrome. It also is applied to the skin to treat rashes and warts. The plant contains several toxic alkaloids and may be associated with liver damage.

9. Green Tea Extract Powder (camellia sinensis)

This one is a little tricky. Green tea contains polyphenols, including one called catechin which works as an antioxidant. Green tea has been used for weight loss, cancer prevention, alertness, depression, headaches, and Parkinson's disease, just to name a few. It has also been used topically for genital warts and to soothe sunburn. Green tea is probably effective for genital warts and to lower cholesterol. It is possibly effective for several other conditions, including Parkinson's disease and prevention of high blood pressure and some cancers. Green tea is also considered to be safe when consumed as a drink in moderate doses. However, green tea contains caffeine and in high doses can lead to nervousness and anxiety, sleep problems, vomiting, diarrhea, irritability, irregular heartbeat, tremor, heartburn, dizziness, ringing in the ears, seizures, confusion, and liver damage. Green tea may also reduce the absorption of iron from food. Consuming very high doses of green tea might be fatal.

10. Kava (Kava Kava, Piper Methysticum)

Kava is a drink or extract made from the leaves of the piper methysticum plant native to the western Pacific islands. Kava is a popular drink in the Pacific used to calm anxiety and stress and to treat insomnia. It has also been used for attention deficit disorders, epilepsy, depression, headaches, and pain. Use of kava has been linked to liver damage and death.

11. Lobelia (Asthma Weed, Vomit wort)

Lobelia is a type of flowering plant found in warm climates. It is used for breathing problems such as cough, asthma, bronchitis, and shortness of breath in newborns. It can also be used topically for muscle pain, bruises, insect bites, and poison ivy. This herb is associated with nausea, vomiting, diarrhea, dizziness, tremors, confusion, coma, and possibly death.

12. Pennyroyal Oil (Mentha Pulegium)

Pennyroyal is a type of flowering plant that has been used for centuries to eliminate pests and to induce abortion. In addition to the uses mentioned, this herb has been used for cold relief, cough, fever, headache, and indigestion. This herb is toxic and has caused liver and kidney failure, seizures, and death.

13. Red Yeast Rice (Monascus purpureus)

This product is made from yeast that is grown on rice. Red Yeast Rice is another of the "mixed bag" supplements. It is effective for lowering cholesterol because it contains a substance called monocolin which is similar or identical to lovastatin (an FDA approved drug for the treatment of high cholesterol). The FDA banned red yeast rice for several years, but products without monocolin have reappeared in the last ten years. This product may contain citrinin, a poison that may cause kidney damage. Like "statin" cholesterol lowering drugs, red yeast rice may also cause liver and muscle damage. This supplement should not be taken with statin drugs.

14. Usnic Acid (Beard moss, Tree moss)

Usnic acid is a lichen that grows on trees. A lichen is a combination of algae and fungus that grow together for a mutual benefit. This supplement can be used for weight loss, fever, pain relief, wound healing, and as a cough expectorant. It can also be used for inflammation of the throat and mouth. Usnic acid is associated with liver damage and should be avoided.

15. Yohimbine

As previously mentioned, yohimbine is used to treat erectile dysfunction and low libido. It has also been used to treat anxiety, depression, and obesity. Yohimbine can cause serious side effects such as high blood pressure, seizures, rapid heart rate and rhythm disturbances, kidney failure, heart attack, and possibly death.

Whatever Happened to Sam?

Sam's initial body mass index was 41.1, which put him into the "obese" category. Sam was rightfully concerned about his diet. The types of foods that he had been eating are high in fat, refined carbohydrates, and salt. Sam suffers from high blood pressure, but his blood pressure might be better controlled with weight loss. Caffeine powder is not recommended in people with high blood pressure.

Sam was motivated to lose weight. He was able to lose forty pounds in a year by bringing healthy meals with him on the road. Sam also made better choices when eating out. He traded in his burgers and pizza for salads with lean protein. He also limited his alcohol consumption. Sam's blood pressure has decreased, and he no longer needs medication.

Bottom Line

According to the Dietary Guidelines for Americans 2015–2020, about half of adults in the U.S. have at least one preventable chronic disease. Many of these diseases are related to poor eating practices and lack of physical activity. Also, more than two thirds of adults (and almost one third of children) are overweight or obese. The bottom line is that nutrition is a huge factor in our health. As mentioned above, a healthy eating pattern includes whole fruits, a variety of vegetables, whole grains, low fat dairy products, healthy oils, and a variety of lean protein. In addition, we should limit saturated and trans fat, sodium, and sugar. Supplements may be beneficial for some individuals, but you should check with your health care provider before taking any supplements.

CHAPTER 14

Useful Resources

Chapter 1: Anatomy and Physiology Down There

- *The A.D.A.M. Medical Encyclopedia.* Atlanta: Ebix, Inc. https://medlineplus.gov/encyclopedia.html
- Netter, Frank H. *Atlas of Human Anatomy.* Philadelphia, PA: W. B. Saunders Co., 2014.
- "The Male Reproductive System." WebMD. http://www .webmd.com/sex-relationships/guide/male-reproductive -system.

Chapter 2: Benign Prostate Conditions—When That Walnut is Acting a Little Nutty

- "What is Benign Prostatic Hyperplasia (BPH)?" Urology Care Foundation. http://www.urologyhealth.org/urologic -conditions/benign-prostatic-hyperplasia-(bph).
- "Management of Benign Prostatic Hyperplasi (BPH)." American Urological Association. http://www.auanet.org /guidelines/benign-prostatic-hyperplasia-(2010-reviewed -and-validity-confirmed-2014).

Chapter 3: Prostate Cancer– Fingering the Little Culprit

- "Prostate Cancer." American Cancer Society. https://www .cancer.org/cancer/prostate-cancer.html.
- "National Comprehensive Cancer Network." NCCN Guidelines for Patients. Accessed August 31, 2017. https://www.nccn .org/patients/guidelines/cancers.aspx.
- "ProstAware." http://www.ProstAware.org

Chapter 4: Erectile Dysfunction/Impotence—When "Mr. Happy" Goes to Sleep or Developing a Weapon of Mass Destruction

- National Kidney and Urologic Diseases Information Clearinghouse
- National Institute of Diabetes and Digestive and Kidney Diseases
- American Urological Association
- American Diabetes Association
- "American Association of Sex Educators, Counselors, and Therapists." http:// www.aasect.org
- "ED Guidance." http://edguidance.com/
- "Vacuum Erection Devices (VED): Basic Principles." Advanced Urological Care PC. https://www. urologicalcare.com/erectile-dysfunction/vacuum -devices-ed-treatment/
- Montague DK, et al, for the AUA Guidelines Panel on Erectile Dysfunction. *The Journal of Urology.* 1996; 156: 2007–2011.
- Donatucci CF. In: Mulcahy JJ, ed. *Male Sexual Function.* Totowa, NJ: Humana Press Inc; 2001: 253–261.
- Levine LA, Dimitriou RJ. *Urologic Clinics of North America.* 2001; 28: 335–341.

- "Erectile Dysfunction: Overview." Advanced Urological Care PC. https://www.urologicalcare.com/erectile-dysfunction/ed-overview/

Chapter 5: EJD or Ejaculatory Dysfunction—The Other Major Sex Problem "Hold on, I'm Cumin'!"

- Michael E. Metz PhD and Barry W. McCarthy PhD *Coping With Premature Ejaculation: How to Overcome PE, Please Your Partner & Have Great Sex*. Oakland: New Harbinger Publications, 2003.
- "Last Longer in Bed: Your Guide to Overcoming Premature Ejaculation." *Men's Health Magazine.*
- Nat Eliason. *Come Again? What Men Should Know About Amazing Sex*. Provo: Kiserson Media., 2017.

Chapter 6: Vasectomy—The Prime Cut

- "Vasectomy: What You Should Know." WebMD. http://www.webmd.com/sex/birth-control/vasectomy-14387
- Hallie Levine. "6 Things Every Woman Needs To Know About Vasectomy." Prevention. http://www.prevention.com/sex/what-you-need-to-know-about-vasectomies
- Urology Care Foundation. http://www.urologyhealth.org/urologic-conditions/vasectomy
- "Answers from experts on vasectomy and vasectomy reversal. "Vasectomy.com. http://www.vasectomy.com
- Sperm Bank Directory. http://www.spermbankdirectory.com.
- "Find Local Sperm Banks With Our New Sperm Bank Directory." Fertility Nation. http://www.fertilitynation.com/find-local-us-sperm-banks-with-our-new-sperm-bank-directory/

Chapter 7: Battling Low Testosterone—When the Grapes Turn to Raisins

- Christopher P. Steidle and Janet Casperson. *Sex and the heart: erectile dysfunctions link to cardiovascular disease.* Omaha, Neb.: Addicus Books, 2008.
- Abraham Morgentaler. *Testosterone for life: recharge your vitality, sex drive, muscle mass & overall health!* New York: McGraw-Hill, 2009.
- American Society of Andrology
- International Society of Andrology

Chapter 8: Problems in The Pouch—When the Family Jewels Don't Shine

- "What is Male Infertility?" Urology Care Foundation. http://www.urologyhealth.org/urologic-conditions/male-infertility.
- "What is Testicular Cancer? "Urology Care Foundation. http://www.urologyhealth.org/urologic-conditions/testicular-cancer.
- "What are Undescended Testicles (Cryptorchidism)?" Urology Care Foundation. Accessed August 31, 2017. http://www.urologyhealth.org/urologic-conditions/cryptorchidism.

Chapter 9: Osteoporosis in Men—We Don't Like Soft Bones

- National Osteoporosis Foundation, http://www.nof.org
- National Institute of Health (NIH), www.bones.nih.gov
- Osteoporosis and Related Bone Diseases National Resource Center, https://www.niams.nih.gov/Health_Info/Bone/default.asp
- International Osteoporosis Foundation, https://www.iofbonehealth.org

- National Bone Health Alliance, http://www.2million2many
.org

Chapter 10: When There's a Pain in the Pelvis That Won't Go Away—When It Really Hurts Down There

- "Interstitial Cystitis/Painful Bladder Syndrome." Interstitial Cystitis Association. http://www.ichelp.org/men.
- "Real Men Get IC." Interstitial Cystitis Association. https://www.ichelp.org/wp-content/uploads/2015/06/IC-Men-Brochure.pdf
- Hans C. Arora and Daniel A. Shoskes. HC. "The enigma of men with interstitial cystitis/bladder pain syndrome." *Translational Andrology and Urology*. 2015 Dec; 4(6): 668–676. https://www.ncbi.nlm.nih.gov/pmc/articles/PMC4708534/
- CP Smith. "Male chronic pelvic pain." *Indian Journal of Urology*. 2016 Jan-Mar; 32(1): 34–39.
- "Chronic Prostatitis / Chronic Pelvic Pain Syndrome." Male Pelvic Floor. http://malepelvicfloor.com/cpps.html
- JC Nickel. "Management of Men Diagnosed With Chronic Prostatitis/Chronic Pelvic Pain Syndrome Who Have Failed Traditional Management." *Reviews in Urology*. 2007 Spring; 9(2): 63–72
- David Wise D and Rodney Anderson. *A Headache in the Pelvis*. San Francisco: National Center for Pelvic Pain Research, 2006

Chapter 11: Incontinence: Diapers—You Don't Have to Depend on Depends™

- The National Association For Continence, https://www.nafc
.org
- "Kegel exercises for men." Tulane Urology Garden District. http://neilbaum.com/articles/kegel-exercises-for-men-non-medical-treatment-of-overactive-bladder-2

- "Urinary incontinence in men." WebMD. http://www.webmd .com/urinary-incontinence-oab/tc/urinary-incontinence -in-men-topic-overview#1
- "Urinary incontinence." Mayo Clinic. http://www.mayoclinic .org/diseases-conditions/urinary-incontinence/basics/ causes/con-20037883
- Neil Baum, "Men You Don't Have To Depend on Depends™ After Prostate Surgery." Dr. Neil Baum's Urology Blog. (blog), https://neilbaum.wordpress.com/2015/10/24/men-you-dont -have-to-depend-on-depends-after-prostate-surgery/

Chapter 12: Sexually Transmitted Infections—An Ouch in the Crouch

- "Sexually Transmitted Diseases (STDs)." Centers for Disease Control and Prevention. July 18, 2017. Accessed August 31, 2017. https://www.cdc.gov/std/default.htm.

Chapter 13: Diet and Nutritional Supplements for "Down There" Health

- Academy of Nutrition and Dietetics, www.eatright.org
- U.S. Department of Agriculture (USDA), https://www .nutrition.gov
- U.S. Food and Drug Administration, https://www.fda.gov /Food/DietarySupplements/default.htm
- National Institute of Health Office of Dietary Supplements, https://ods.od.nih.gov
- United States Department of Agriculture (USDA), https: //www.choosemyplate.gov
- Office of Disease Prevention and Health Promotion, https: //health.gov/dietaryguidelines/

Conclusion

Nearly 100 percent of men between the ages of forty and seventy-five have issues and challenges "down there" at some time in their lives. You should not suffer in silence. We have written this book because many men are uncomfortable discussing these issues with their friends, their partners, and even with their doctors. The truth is that the unique male organs between your belly button and your knees can be the most celebrated area of your body—and they should be.

This book represents a compilation of stories from men just like you—men who have experienced similar challenges and uncertainties "down there." We want you to recognize that help is available. Your road to recovery begins with understanding your problem, and the underlying symptoms, pain, or discomfort you are experiencing. The next step is to speak to your doctor, who will then craft a plan of action or refer you to a male health specialist who can help you.

You must take responsibility for executing your personal plan of action. Remember, to know and not to do, is not to know. You must become a partner with your doctor and follow the course of action that is laid out for you. Only then will you gain control of the issues with your tissues

This book should serve as guide to help make you a better patient. Hopefully, you will read this book before your appointment with your doctor. Now, you are in a position to understand your doctor's advice—and you have the knowledge necessary to help your doctor reach an accurate diagnosis and start you on the road to recovery.

We want to empower you to develop a support system with which you can discuss what is going on "down there"—whether it's your physician, your partner, your friends, or your relatives. It's important to feel comfortable and confident sharing your troubles and triumphs, seeking advice from those who have been on the "down there" road before, and sharing your experiences with those who are still striving to understand what is going on "down there."

Finally, we would like to hear from you. If you have a success story and would be willing to share it with your fellow men in an anonymous fashion, please share it with us so we can help others learn from your experience. Also, if you are grappling with a challenge, perhaps we can connect you to someone who can help. You can reach us at doctorwhiz@gmail.com for Dr. Neil Baum and ScottDMillerMD.com for Dr. Scott Miller.

Index

About the Authors

Neil H. Baum, MD, is Professor of Clinical Urology at Tulane Medical School in New Orleans, Louisiana.

Dr. Baum has written over 250 peer-reviewed articles and seven books on men and women's healthcare issues. He is is the author of *ECNETOPMI-Impotence It's Reversible* and *What's Going on Down There-Improve Your Pelvic Health*, for women.

Dr. Baum was the columnist for *American Medical News* for more than twenty-five years. Dr. also Baum wrote the popular column, "The Bottom Line," for *Urology Times* for more than twenty years. He also focuses on patient education and has written extensively on enhancing communication with patients and their families.

Dr. Baum has made an effort to simplify the communication with patients who have concerns about prostate cancer screening, erectile dysfunction, and urinary incontinence in men.

Dr. Baum is married to Linda, has three children, Alisa, Lauren, and Craig, and two grandchildren, Ryan and Shelby.

Scott D. Miller, MD, is a men's health expert who believes patients should be informed and involved in their care. He has dedicated his career to educating the community and his colleagues through original writings and regular television and radio appearances. As a board-certified urologist for over twenty years, Dr. Miller has one of the largest, most diverse experiences in laparoscopic and robotic urology in the Southeast. He is an advocate for minimally-invasive techniques so that patients can quickly return to their normal routine.

Dr. Miller practices at WellStar Medical Group and is medical director of robotic surgery at WellStar North Fulton Hospital. He is founder and president of ProstAware, a non-profit prostate cancer awareness organization. He is also an accomplished musician and composer who has written a variety of songs, including one about the emotional impact of a cancer diagnosis. He lives in Atlanta, Georgia with his wife Mindi, who contributed two chapters to *How's It Hanging?*